T0136721

URANUS AND NEPTUNE

KOSMOS

A series exploring our expanding knowledge of the cosmos through science and technology and investigating historical, contemporary and future developments as well as providing guidance for all those interested in astronomy.

Series Editor: Peter Morris

Already published:

Asteroids Clifford J. Cunningham
The Greatest Adventure Colin Burgess
Jupiter William Sheehan and Thomas Hockey
Mars Stephen James O'Meara
Mercury William Sheehan
The Moon Bill Leatherbarrow
Saturn William Sheehan
Soviets in Space Colin Burgess
The Sun Leon Golub and Jay M. Pasachoff
Uranus and Neptune Carolyn Kennett
Venus William Sheehan and Sanjay Shridhar Limaye

Uranus and Neptune

Carolyn Kennett

REAKTION BOOKS

To Charlie, who made this possible

Published by Reaktion Books Ltd
Unit 32, Waterside
44–48 Wharf Road
London N1 7UX, UK
www.reaktionbooks.co.uk

First published 2022

Printed and bound in India by Replika Press Pvt. Ltd

A catalogue record for this book is available from the British Library

ISBN 978 1 78914 641 7

CONTENTS

Preface

This is a book about Uranus and Neptune, the two giant planets that orbit the outer solar system. Predominantly formed of gas, they are significantly smaller in size than their neighbours, the gas giants Jupiter and Saturn. The differences continue: they are primarily composed of heavier elements, rather than hydrogen and helium, which make up 90 per cent of the mass of the larger gas giants. It is these heavier elements that give Uranus and Neptune their names, as they are collectively called ice giants owing to the 'ices' contained in their mantles. Chemical compounds, water, ammonia and methane have much higher freezing points than hydrogen and helium and are thought to exist in a solid state within an interior ocean, where it could rain diamonds. Methane in the outer cloud layers of the atmosphere gives the planets their colours: Uranus a blue-green azure and Neptune a deep midnight indigo.

Unknown to humankind until modern times (despite Uranus being a naked-eye object), our understanding of these two outlying planets has increased significantly since they were visited by Voyager 2 in 1986 (Uranus) and in 1989 (Neptune). In the thirty years since, there have been no further missions to the planets; instead, they have been watched from a distance by viewers using powerful telescopes such as the Hubble Space Telescope and the ground-based Keck. Technological advances in the astronomical capabilities of these telescopes have pushed the boundaries of what is possible

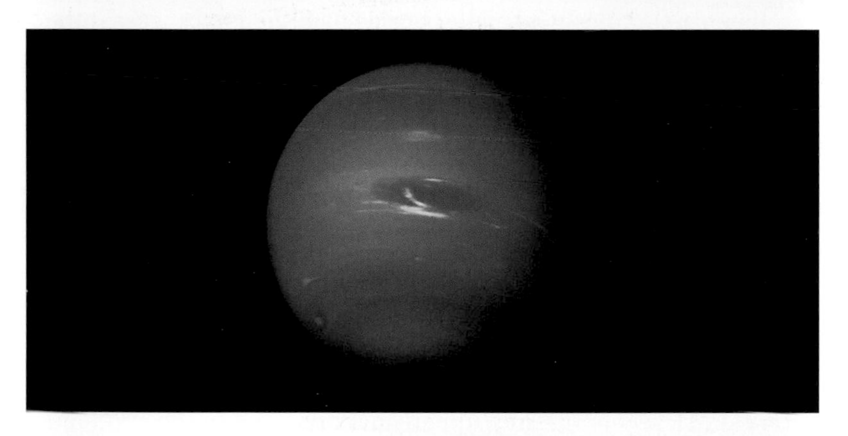

Uranus and Neptune as imaged by Voyager 2.

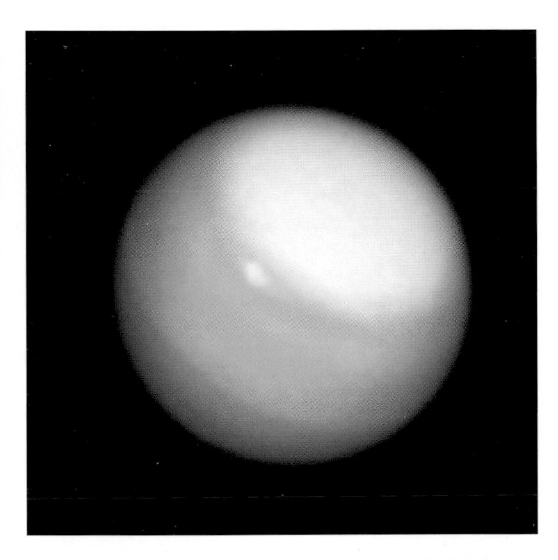

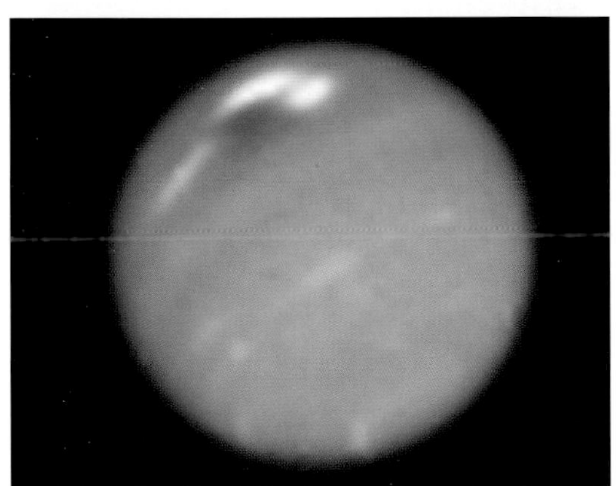

Uranus (top) and Neptune imaged by the Hubble Space Telescope in 2018.

to observe at these great distances, leading to numerous discoveries. Since Voyager has continued on its journey into the outer solar system, dark spots have been observed forming and disappearing on Neptune, with a greater frequency than the ever-present Great Red Spot on Jupiter.

Voyager photographed a placid uniform surface on Uranus, which was a disappointment to the team of scientists and the waiting public. The surface is now known to be more dynamic and, since the flypast, its white polar hood has moved from the south to the north pole. This process was set in motion by the planet moving past equinox, in 2007, where the south pole dipped into permanent darkness, while the north came under the enduring solar glare of summer. Seasons last for 21 years on Uranus, created by the planet's bowling-ball orientation. Its highly unique axial tilt of nearly 98° to the plane of its orbit is a reminder of a past cataclysmic collision during its early history.

Although our understanding of these planets has greatly increased, due to the incredible distances involved there are many outstanding questions. The interior structure of the planets, their unusual magnetospheres and the formation and structure of the ring systems all hold mysteries. Historic observations of ring systems by skilful astronomers (Uranus by William Herschel and Neptune by William Lassell) proved to be false trails and it would take the occultation events, such as that of the star SAO 158687 by Uranus, for the first rings to be identified in 1977, whereas Neptune would have to wait for the

Voyager flypast. Currently thirteen distinct rings are known around Uranus and five principal rings around Neptune. The Adams ring, the outermost known ring around Neptune, contains unusual denser areas called arcs. The longevity of the arcs is unknown and changes have occurred since Voyager imaged them in detail; although the arcs are similar to Saturn's G ring, they are little understood.

A number of orbiting moons have the potential to harbour life, particularly the giant Triton, a captured Kuiper Belt object orbiting Neptune, which is similar in size but slightly larger than the dwarf planet Pluto. There is a real possibility that more moons orbit the planets; the James Webb Telescope will have the capability to look further and deeper with more clarity than ever before and it is expected that more moons will emerge from the dark depths. A return to the planets via a dedicated space mission would offer opportunities to gain a greater understanding about their features and their origins. The lack of space missions to Uranus and Neptune is further highlighted by the recognition that ice giants are currently the largest-known class of planet in the universe. This book will consider both planets in turn, Uranus and then Neptune, but first it will discuss the origins of the two ice giants and the circumstances surrounding their formation.

ONE

Uranus, Its Discovery and Pre-Voyager Observations

Uranus is special for many reasons: principally, because it is the first planet to be discovered within our solar system in modern times. Since humans started to watch and understand the motions of the heavens, Saturn had been considered the outermost planet. This was all to change when Uranus was identified as a planet, effectively doubling the size of the solar system, although it had been hiding in plain sight, often mistaken for a star. Uranus's story starts with the formation of the solar system.

Origins

A giant molecular cloud of dust started to collapse 4.6 billion years ago, possibly the result of a supernova whose shockwave kickstarted the process. Within the spinning disc of gas, a solar nebula was born. This swirl of gas would draw material into its centre. Here the heat and pressure would become so intense that eventually a star would ignite and much of the remaining material in the disc would come together to form the planets we see today.

Scientists still have much to learn about the first years of our solar system, when Uranus was formed. One recently discovered object, which has shed light on the formation of the planets, is called Arrokoth. The object is located in the Kuiper Belt, a region outside the orbit of Neptune, which is more than 6 billion km from the Sun. New Horizons, a NASA space probe, flew past this pristine remnant in 2014 and the images beamed back to Earth were surprising. The team found evidence of planet formation in action. When scientists scrutinized the images, Arrokoth looked like two spherical objects that had attached themselves together. At first it was anticipated that they were two spheres, but when the spacecraft flew past, it became evident they were flat, more like a pancake in shape. The shape suggests the two halves of the object had formed separately before finally merging.

Previous theories had suggested that formation would have happened in a dramatic way, with the primordial disc containing particles that crashed into each other to form small objects and these, in turn, having their own cataclysmic comings together, building up into larger bodies. Examinations of Arrokoth show that a cataclysmic merging could not be the case. The structure had no signs of stress and no fractures or flattening, which would have been noticeable from a violent crashing together of the two halves. It is now considered that objects would have come together through a gentle clumping; formation would have happened through a much gentler dance, with the bodies being squashed together in a slow manner, through the attractive force of gravity. They would have gradually swept up material within their orbits. Perhaps, instead of forming in chaos, Uranus was made by building blocks of material clumping together in a more serene manner. In comparison, the planet's journey to its present orbit has been nothing less than exciting and destructive. This is demonstrated by its tilt, the rotational axis of which is nearly on the same plane as the planet's orbit. The early solar system would have been crowded, containing a larger number of planetary bodies than exist today. Uranus was hit by another similar-sized object. This cataclysmic event not only tilted the planet but created many of its smaller moons.

It is highly unlikely that Uranus formed in its current orbit, along with its neighbour Neptune. During these early years both planets underwent migration. This is most likely due to the outwards migration of the large planets Jupiter and Saturn, whose gravitational influences dictate the region. It is possible that Neptune and Uranus swapped places and, initially, Neptune formed closer to the Sun than Uranus. The solar system is still dynamic, but it is a much quieter place than in those first 100 million years, and the planets Uranus and Neptune have now settled into orbits that are little changed from the time of their discovery. Uranus and Neptune's similarities suggest they had a common background. Along with their colours, the two planets have a similar mass, size and composition. Both of them are much smaller in size than the two other gas giants, Jupiter and Saturn, whose size and bulk dominate our planetary system. In the early days of the solar system their paths deviated and Uranus is now considered to be an atypical ice giant and Neptune an archetype and far more representative of this class of planet; the reasoning behind this will be explored more throughout the book.

Discovering a Planet

When Uranus was discovered in 1781 by William Herschel (1738–1822) it was such a surprise that it took two years for the object to be officially recognized as a planet. Herschel made this groundbreaking astronomical discovery through his 'seven-foot' reflector (6.2-inch aperture; 7 ft is the focal length) in his back garden on 13 March 1781. He was a German musician who lived in Bath with his sister, the astronomer Caroline Herschel (1750–1848). William had a lifelong passion for astronomy and, although considered at this point an amateur, his skills in mirror making were among the best in the world. He had already conducted a comprehensive review of the night sky, cataloguing stars he could

William Herschel, *c.* 1795.

observe down to an apparent magnitude of 6, and had started working on a second review which would examine all the stars down to an apparent magnitude of 8.

William Herschel had recently moved house in Bath from River Street to 19 New King Street. The Herschel household had previously resided there and they had liked the property, as the lower floor could be used as a workshop to make the mirrors for telescopes. On the night of 13 March 1781, Caroline was still clearing their previous residence, and William was observing alone. He set up his telescope in the south-facing rear garden and continued with his review of the night sky.

The entries in his book for that night read:

> Pollux is followed by 3 small stars at about 2′ and 3′.
> Mars as usual.
> In the quartile near Zeta Tauri the lowest of two is a curious either nebulous star or perhaps a comet.
> A small star follows the Comet at 2/3rds of the field's distance.

Looking through the eyepiece, he paused on what he described as 'a curious either nebulous star or perhaps a comet'. When the object appeared in the eyepiece as a round disc, it was noticeably larger than the other stars in the sky. William immediately changed his eyepiece and increased the magnification of his telescope from 460× to 932× . He then carefully measured its position with a micrometre. When he returned to the same spot a few hours later it was clear the object had moved. This was definitely no star, nor was it a nebula, but was presumably a comet. Excited by the thought of this discovery, William Herschel was quick to present his account of the 'comet' at the Bath Philosophical Society on 22 March 1781. William Watson (1744–1824), a close friend, communicated the discovery to the Royal Society on 26 April 1781:

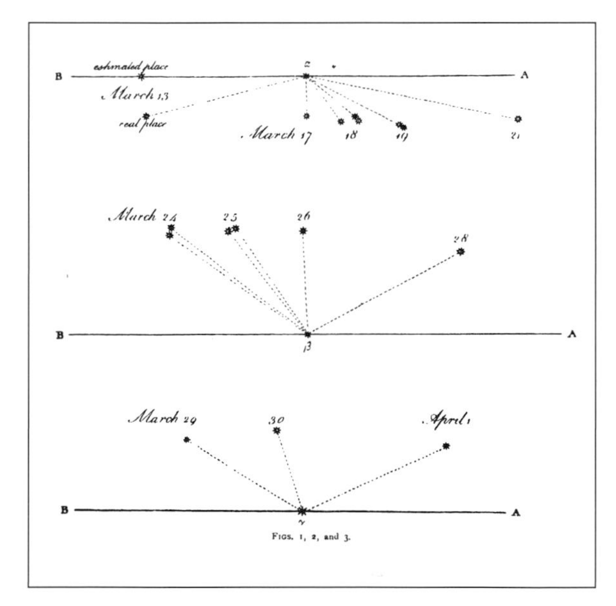

Successive places of the new planet from 13 March to 28 March 1781 in relation to stars α = BD +23°1069, β = BD +23°1074. Taken from Herschel's paper to the Royal Society, 26 April 1781.

Replica of William Herschel's reflecting telescope. Made in mahogany wood, the octagonal barrel is supported on a mount that allows it to be moved in elevation.

> While examining the small stars in the neighbourhood of H Geminorum [1 Geminorum], I perceived one that appeared visibly larger than the rest; being struck with its uncommon magnitude, I compared it to H Geminorum and the small star in the quartile between Auriga and Gemini, and finding it so much larger than either of them, suspected it to be a comet.[1]

From here it was not long before astronomers far and wide had heard about this new celestial object.

Professional astronomers such as Nevil Maskelyne (1732–1811), the Astronomer Royal, initially struggled to find the object. William Herschel wrote on 6 April that it did not appear like a comet, as visually it was 'perfectly sharp upon the edges, and extremely well defined, without the least appearance of any beard or tail'.[2] Now that the object was being watched across Europe, the astronomical community could turn its attention to describing its orbit, a task which was both laborious and time consuming. William Herschel made a number of observations, which he believed would show the 'comet' he had first seen in March 1781 approach the inner solar system. In effect, though, it was doing the opposite; it was receding.

Calculating the orbit of the object proved tricky, as it was not following a predictable regular path through the sky. A number of astronomers, including Pierre Méchain (1744–1804) in Paris and Anders Lexell (1740–1784) in Russia, attempted to work out a perihelion distance from the Sun. Méchain reported a perihelion distance of 0.46 astronomical units (AU) on 23 May 1781, while Lexell stated a perihelion distance of 16 AU on 10 April 1789. Maskelyne wrote, on 4 April 1781, that it was 'very different from any comet I ever read any description of or saw' and he expanded this on 23 April 1781, saying, 'as likely to be a regular planet moving in an orbit very nearly circular around the sun as a comet moving in a very eccentric ellipsis'.[3]

The news was remarkable but it would still take some months for the planet to be officially confirmed, and for a while the astronomical community called the object a 'new star'. The significance of the discovery at first seemed to be lost on William Herschel himself, who said,

> With regard to the new star I may still observe that tho' we are not sufficiently acquainted with its nature, yet enough has been seen already to shew that it differs in many essential particulars from Comets and rather resembles the condition of Planets.[4]

This is a defining moment in astronomy. William Herschel's planet was the first to be discovered since ancient times. The rewards were great and he received fame and wealth, including a pension from George III.

Naming a Planet

Naming a new planet was never going to be a straightforward affair. This was the first time that a name would be given to a planetary body in the solar system since antiquity and giving it a name of suitable standing would be of the upmost importance in the minds of the astronomers and public alike.

Many wanted the planet to be called 'Herschel' after the hero of the hour, but he was not happy to take the honour and was given the opportunity to suggest his own name for the planet. William Herschel decided the planet should be called 'Georgium Sidus', after the king of Great Britain and elector of Hanover, George III, who became Herschel's patron. This did not sit well with astronomers from further afield, not only because it glorified his patron but because *sidus* meant star and not planet. William Herschel tried to modify Georgium Sidus by changing it to 'Georgian Planet' or 'Georgian', but the name was not popular or used commonly abroad.[5] A number of mythical names were suggested, including 'Cybele', 'Austräa' and 'Neptune'. Johann Elert Bode (1747–1826)

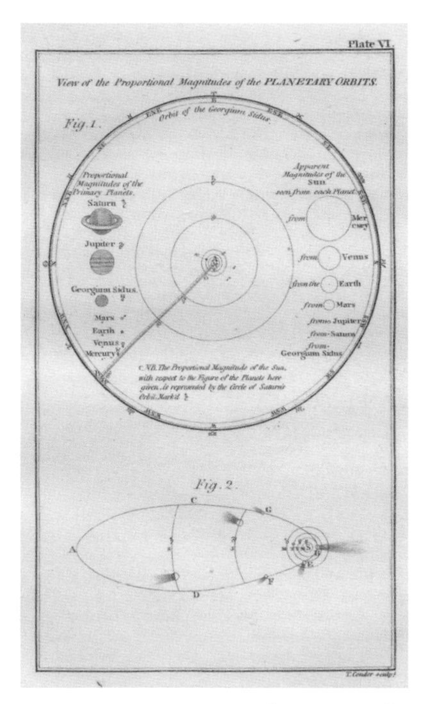

Engraved diagram of the planetary orbits in the solar system, including that of Georgium Sidus (Uranus), from Margaret Bryan, *Compendious System of Astronomy* (1797).

proposed the name 'Uranus' after the Greek god of the sky. This soon found favour across Europe, being championed by Abbe Maximilian Hell (1720–1792). Bode and Hell were eminent European astronomers and their views held great weight, although the British were not initially in accord. Eventually the name Uranus prevailed but it would take over seventy years to do so, with both Herschel and particularly Georgium Sidus being used regularly on star maps and in books in Britain until well into the 1840s.[6]

The pronunciation of Uranus leads to much discussion. The 1920s *Oxford English Dictionary* suggested 'you-ran-us' – Sir Patrick Moore, when presenting *The Sky at Night* television programmes, would pronounce the word in the same manner. The alternative and more phonetic pronunciation that softens 'ur' and then enunciates 'anus', which has caused much mirth, became popular by the 1980s when the *Merriam-Webster Dictionary* referred to this version as the correct way to articulate the word. The second pronunciation fell out of favour in 1986 when Voyager was due to arrive at the planet, as newscasters were concerned about detracting from the science surrounding the planetary mission with the use of this version.

Missed Opportunity

Bode now turned his attention to the past records to see how many occasions it had been observed previously. He anticipated this would help him to calculate the orbit of the new planet. Bode examined the major star catalogues of that time, including those made by the Danish astronomer Tycho Brahe and John Flamsteed.

He quickly found that the planet had indeed been missed. One star stood out in Flamsteed's catalogue, numbered 34 Tauri.

Working backwards it would have been in the same area as Uranus when it was catalogued on 23 December 1690. Bode realized that this 'star' could no longer be found and Flamsteed's 34 Tauri was in the position that Uranus had been in 1690. Looking further into Flamsteed's records it is now known that he recorded it on at least five other occasions: once in 1712, possibly in 1714 (this may have been his assistant Joseph Crosthwaite at the telescope) and then four times in 1715 between 4 March and 29 April. On all these occasions Flamsteed had catalogued it as a star.

Pierre-Charles Le Monnier (1715–1799) had also misidentified the object as a star. His observations showed he had incorrectly catalogued it on thirteen separate occasions. Le Monnier worked in Paris as an astronomer and a professor of physics. He was considered to be a difficult character who found it hard to make friends. His work was messy and his records were often scrawled onto scraps of paper, with badly organized observations and, at times, illegible handwriting. Le Monnier first observed the planet on 14 October 1750. But it was the consecutive observations made across four nights in January 1769 that seem puzzling. Why had he not realized he was watching a body that moved and not a fixed star? With hindsight it appears remarkable that a professional astronomer would not have recognized a planet in their telescope. For Le Monnier it seems that luck was not on his side. Uranus was reaching its 'stationary point' when viewed from Earth and it would have been almost impossible to detect its motion over the course of those evenings. If Uranus had been at a different stage in its orbit, no doubt he would have discovered the planet more than a decade before William Herschel. To add salt to the wound, Le Monnier would live until 1799, realizing his missed opportunity and that he had indeed observed the planet on a number of occasions, as three of his pre-discovery observations would be identified by the famous mathematician Pierre-Simon Laplace (1749–1827) in 1788; Le Monnier's further sightings would be identified by astronomer Alexis Bouvard (1767–1843) in 1820.

It is now thought that Uranus had been recorded on 23 occasions by five different astronomers before its discovery as a planet by William Herschel. There were three observations made by James Bradley (1692–1762) and one by Johann Tobias Mayer (1723–1762).[7] In addition Kevin Kilburn has identified a possible observation made by John Bevis (1695–1771) at the end of June 1738. The pre-discovery observations of Uranus are listed below, compiled from a list by William Sheehan:[8]

Date of Observation	By Whom
23 December 1690	John Flamsteed
2 April 1712	John Flamsteed
1714	John Flamsteed (perhaps his assistant, Joseph Crosthwaite)
4 March 1715	John Flamsteed
5 March 1715	John Flamsteed
10 March 1715	John Flamsteed
29 April 1715	John Flamsteed (could be 1712)
1748	James Bradley
1750	James Bradley
1750	Pierre-Charles Le Monnier
14 October 1750	Pierre-Charles Le Monnier
3 December 1753	James Bradley
26 September 1756	Johann Tobias Mayer
15 January 1764	Pierre-Charles Le Monnier
27 December 1768	Pierre-Charles Le Monnier
30 December 1768	Pierre-Charles Le Monnier
15 January 1769	Pierre-Charles Le Monnier
16 January 1769	Pierre-Charles Le Monnier
20–23 January 1769	Pierre-Charles Le Monnier (four nights)
18 December 1771	Pierre-Charles Le Monnier

Uranus had now been identified as a planet and its orbit was being calculated; astronomers were determined not to lose sight of it again. In the depths of the outer solar system, it orbited almost twice the distance from the Sun of its nearest neighbour, Saturn. The orbital period of 84.2 years was first calculated by Laplace. This positioned the planet at a mean distance of 20 AU, which fitted well with the so-called Titius–Bode law, named after Johann Daniel Titius (1729–1796) and Johann Elert Bode (1747–1826). This empirical 'law' suggests that each successive planet would orbit roughly twice the mean distance from the Sun as the planet that preceded it. It was based on observations of the planets from Mercury to Saturn, but the law no longer holds for the outer planets and it is now just a numerical curiosity. Taking Saturn's mean distance of 9.5 AU, Uranus seemed to fit the pattern at 20 AU. At first glance the planet appeared to be in perfect harmony with what was understood about the rest of the solar system.

As astronomers monitored Uranus, they found its motion was not following the path they had predicted, and their recent observations did not tie in with the historic observations, particularly the early one of the misidentified Tauri 34, made by Flamsteed in 1690. The astronomer Alexis Bouvard discarded all the observations made before the discovery, as he felt there were errors in the observations and they could not explain what was happening to the orbit of Uranus. He devised a new set of tables made up of observations from the date of discovery to 1821 and calculated a new orbit. Although he was initially happy with the result, this was not the end of the story, as Uranus was soon wandering away again from its predicted orbital path. There was one obvious answer: the planet was being pulled by another body outside its orbit. Something was acting gravitationally on Uranus and creating the apparent errors that had been observed. Uranus was a naked-eye object but was not discovered until 1781. Finding another planet even further away from the Sun would be an enormously difficult task. This problem was taken up by two mathematicians, John Couch Adams and Urbain-J.-J. Le Verrier, independently in the 1840s, as we will see in Chapter Five.

Uranus as a Planet

Astronomers were keen to find out more about the new planet. Just how massive was it? What was its diameter? And were there any blemishes on its bluish-green surface? William Herschel turned his telescope to the planet again and made ten measurements of its diameter between November 1781 and November 1782.[9] At just 3.7 arc seconds, the disc of the planet was very small in his eyepiece. But he used the best equipment he owned and the highest possible magnification. He suggested that Uranus had an equatorial diameter of 55,067 km, just larger than the modern value of 50,724 km. He was also able to observe that the disc had a slight polar flattening; it appeared as an oblate shape rather than a perfect sphere. He put this flattening at about 10 per cent of the diameter; modern measurements now make this as low as 2 per cent.[10]

William Herschel modified his equipment in January 1787, in an effort to make more extensive observations. He removed his flat secondary mirror in the telescope and tilted the speculum; this optical configuration is now known as a front-view or Herschelian arrangement. It enabled him to look directly through the eyepiece, removing the reduction in brightness that resulted from the reflection of light off a secondary mirror. He discovered not one but up to six satellites. Only two of these satellites were confirmed by William Lassell (1799–1880) in 1851. These were named 'number one' and 'number two'; they would later be named 'Titania' and 'Oberon' by William's son, John (1792–1871). It is also a possibility that one was Umbriel; the others were questionable or non-existent.

William Herschel watched satellites 'number one' and 'number two' orbit the planet. Completely unexpectedly the satellites were

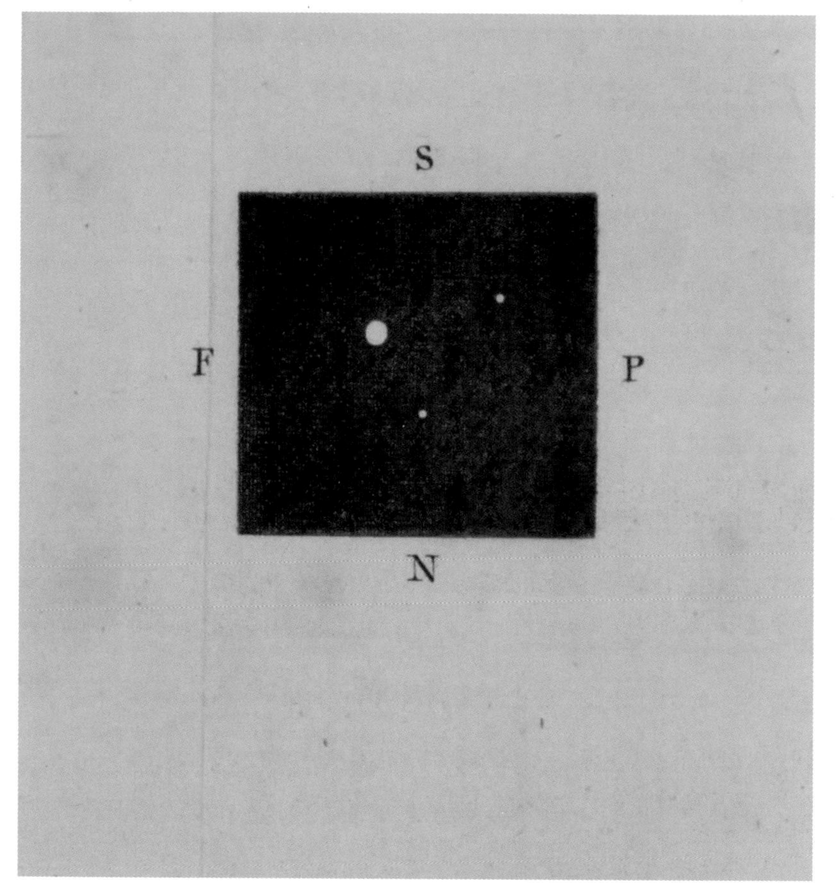

Herschel's drawing of the 'Georgian Planet attended by two Satellites', 11 February 1787, observed through his 18.7-in. reflector.

orbiting around Uranus at an extremely high inclination. If the moons orbited within the planet's equatorial plane, it would mean that the planet was also similarly tilted. Previously, Saturn had the largest axial inclination, at 26°.73. All the other planets in the solar system were tilted even less than this, with Mercury having no tilt at all. By contrast, Uranus's tilt is a remarkable 97°.77.

As well as looking for satellites, William Herschel's work in 1788 enabled him to calculate the mass, volume and density of the planet. He concluded that Uranus was 17.74 Earth masses – larger than the current value of 8.681×10^{25} kg or 14.5 times more massive than Earth. He also measured its density, calculating that it had a density of 1.114 g/cm^3 or 0.22 as dense as the Earth, which is remarkably close to the current value of 1.27 g/cm^3. Only Saturn is less dense than Uranus in our solar system.

Herschel's Mistaken Identity of Rings?

On 4 February 1787, while observing Oberon, one of the satellites he had found orbiting Uranus, William Herschel wrote in his journal, '20 feet reflector, power 500. Well defined; no appearance of any ring.'[11] This is the first mention of rings in relation to the planet. It implies that he may have observed them prior to this date, and on 4 March 1787 he observed two rings, which were perpendicular to each other. However, on 16 March 1787 he recorded only one ring. This confused Herschel and he continued to inspect the planet over the coming months. He started to suspect there were no rings at all. In February 1789 he believed he had seen the ring again, and when he turned his speculum (mirror) around by 90° (enabling him to check for image defects in the mirror, as although the orientation of the object would turn with the mirror, the rings would not) he found that the ring was still in place. His drawings show a flare from one side of the planet. This pattern is often attributed to an image defect called a 'coma'. Herschel was concerned about imperfections within his equipment, as he was working at the boundaries of what was possible at that time. His suspicions would continue over the coming years, and even with a change in mirror and reflector, he remained undecided about his early observations of the ring system. By February 1794 all mentions of a ring were forgotten. The historian Richard Baum believed that Herschel's ring references were a result of optical illusions created by his telescopes, namely astigmatism, or a difference in curvature of the mirror.[12]

There are few recorded observations of rings around Uranus between Herschel's observations and their official discovery in 1977.

Stuart Eves, from Surrey Satellite Technology Ltd, examined the original observations. Herschel observed two rings perpendicular to each other, which cannot be the case.[13] Eves argues that many aspects of the original observation are correct, including a similarity to the ε (Epsilon) ring's relative size to Uranus. Herschel correctly described its colour as red (a feature only recently confirmed by the Keck Telescope, based in Hawaii).[14] The lack of further historical observations could have been due to the orientation of the rings: at that time they had a particular set of circumstances putting them in the perfect position for people here on Earth to observe them. This does not explain the oddity of the observation of two perpendicular rings, and on balance it seems most likely that the observations were a result of Herschel's optical limitations.

Methane

The introduction of spectral analysis in the mid-nineteenth century enabled astronomers to move beyond purely visual observations of astronomical bodies and to examine their chemical constituents. Astronomers could look at the reflected light of a planet and, using the absorption bands in its spectrum, they could work out the planet's chemical composition. Father Angelo Secchi (1818–1878) of the Collegio Romano Observatory and, soon afterwards, William Huggins (1824–1910) were able, independently, to detect dark absorption bands within the spectrum of Uranus between 1869 and 1871. Secchi made his announcement in a letter to the French Académie des sciences, saying, 'The spectroscope continues to reveal to us quite curious things, but what I have just discovered will doubtless appear to you, as it does to me, very singular. It concerns the spectrum of Uranus, which shows a quite unexpected constitution.'[15] The dark lines remained a mystery for another sixty years, when they were finally identified as resulting from methane. This was the first detection of methane on the planet.

Early Observations of Surface Markings

Historically, Uranus's surface was thought to be devoid of any activity, possibly due to its lack of an internal heat source, which would create activity and dynamism at the surface. This does not mean, though, that all observations before Voyager 2 showed a featureless disc. In 1870 and then again in 1872, Thomas Hughes Buffham (1840–1896) expertly sketched Uranus. His observations were made with a 9-inch With–Browning refractor from his home in Earith, Cambridgeshire.[16] The surface disc had a number of large features on it, and Buffham reproduced these by drawing stripes and, in one sketch, two large white spots. Could he have seen such markings on the disc? Arthur Francis O'Donel Alexander (1896–1971), a competent amateur astronomer, felt that these observations 'were probably the first in which fairly definite disk features seem to have been repeatedly seen'.[17] The two large spots on the surface were drawn by Buffham on 25 January 1870 and then again on 27 January 1870. A recent reconstruction of the movement of the markings undertaken by Kevin Bailey has shown the two consecutive observations fit with the current known rotation speed of the planet.[18]

Faint spots were observed by the Italian astronomer Giovanni Schiaparelli (1835–1910) in 1883. He had been making observations of the disc in an effort to find out how much polar flattening the planet had when he noted, 'During the . . . observations, and especially when the air was more tranquil, I have been able to verify that Uranus also has spots and variations in colour on its surface.'[19] Schiaparelli was a competent planetary observer working at the Brera Observatory in Milan. He had made comprehensive drawings of the surface of Mars which showed numerous linear structures called *canali* (channels); this word was incorrectly translated as 'canals' and thus fed into the speculation that life existed on the planet. He felt that his observations of Uranus were limited by the equipment

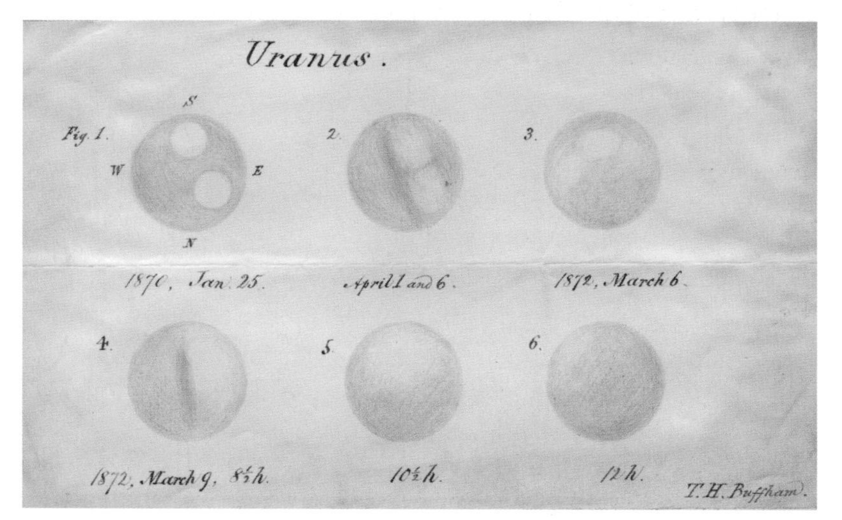

Sketches of Uranus made by Thomas Hughes Buffham and presented at the December 1872 meeting of the Royal Astronomical Society.

he had to hand and considered that the power behind 'great modern telescopes' would ascertain the markings, an idea that would be put to the test later that year by Charles Augustus Young (1834–1908), a professor at Princeton University, using the 23-inch Halsted Observatory refractor. He observed bands that encircled the planet, a lighter one in the equatorial region and darker poles, with the north pole seemingly darker than the south. The banding was faint, although with the correct seeing conditions they became much more distinct. Young likened them to 'the belts of Jupiter viewed with a very small telescope'. He also made a nod to the competency of the Italian observer, saying: 'That Schiaparelli glimpsed these markings with an 8-inch speaks volumes for the Italian observer, instrument and sky, since at Princeton with the 23-inch, they could only be made out in a very vague, faint and far from satisfactory manner.' The belts around the planet also seemed to be inclined at an angle to the plane of the rotating moons. Young had measured them at 32°; other observers had them range from as little as 10° to a much larger 56° by a team at the Paris Observatory. The following year, in 1884, Louis Thollon (1829–1887) and Henri Joseph Anastase Perrotin (1845–1904) at the newly founded Nice Observatory made observations of a dark spot on the surface of the planet. Dark spots would not be confirmed until 2006, more than 120 years later. Thollon and Perrotin observed them in the centre of the disc and compared the dark spots to similar features they had seen on Mars.[20] Perrotin was an excellent observer who discovered six asteroids, including 252 Clementina the following year, but the observations have a question mark hanging over them owing to the lack of further sightings through larger and more powerful telescopes.

An additional observation of bands encircling the planet was made in the early twentieth century by Reginald Lawson Waterfield (1900–1986). Waterfield was a competent observational astronomer who was the Mars section leader (1931–42) and later president of the British Astronomical Association (1954–6).[21] His observing started in 1913 and by 1914 he was using a 250-millimetre refractor owned by James Worthington at the Four Marks Observatory near Winchester.[22] In 1915 he turned the 250-millimetre refractor to Uranus. His resulting sketches depict it in much the same way as Jupiter: he drew a surface covered in broad stripes. It is striking that his observations show the altering tilt of the bands, with the drawings from 1915 showing the bands parallel to the plane of the satellites and a final observation made one year later, in 1916, showing them at a much greater tilt of around 30°.

Over the years, experienced astronomers have questioned observations of markings on the surface of Uranus. 'Even large

Markings on the surface of Uranus drawn by R. L. Waterfield in 1915 and 1916.

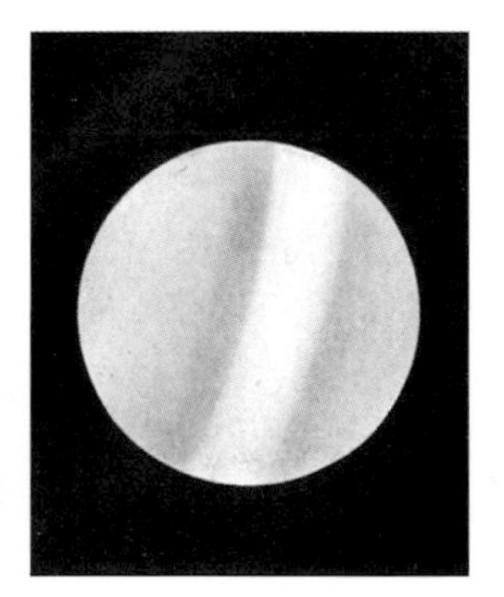

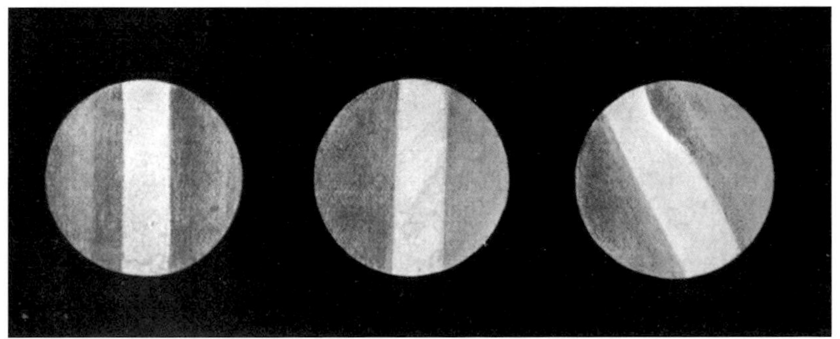

The 24-in. Clark telescope at the Lowell Observatory.

telescopes will show virtually nothing on Uranus's pale disc,' admonished Patrick Moore, who continued, 'I am unhappy about details shown on drawings made by observers with much smaller instruments. Uranus is a very bland world.'[23] However, this negative opinion has been challenged by astronomers recreating the observations made in the past.

Rotation

Because of the lack of features on the surface, it was difficult for astronomers to measure how fast Uranus rotates, so another method had to be found. Early measurements were made at the Lowell Observatory, Arizona, by Percival Lowell (1855–1916) and the brothers Vesto Melvin Slipher (1875–1969) and Earl Carl Slipher (1883–1964), using pioneering spectroscopy techniques. In 1910, using the 24-inch Clark telescope, Vesto Slipher was able to create photographic plates of spectra, which were then measured by Lowell. Using a travelling microscope he was able to determine how fast the lines had shifted across the face of the planet as it rotated, thereby determining the rotation period. In 1912 Lowell reported that this rotation period was 10 hours and 50 minutes.[24] This remained the agreed value for the next 75 years.

A Frigid Temperature

Sir Harold Jeffreys (1891–1989) was a pioneering British planetary scientist. Born in a small mining village in Durham, and a keen naturalist, Jeffreys's mathematical abilities enabled him to attend the University of Cambridge. He was one of a small group of people in the early twentieth century working on planetary structure. His ability to understand the natural world allowed him to have unique insights into the physical properties of the planets. In 1923 he considered the composition of all the gas giants. Until that time, there had been a consensus that these large planets were hot gaseous bodies. Instead, his models showed that the planets were much colder and made up of material with a much lower density than the terrestrial planets. He correctly argued that the two ice giants had surface temperatures lower than 153 K (the current value for Uranus is 59 K) and suggested that a planet's temperature was largely determined by its distance from the Sun. Uranus has little internal heat remaining from its formation, thus any heat found at this frozen planet is received from the distant Sun.

Uranus has a particular set of circumstances, as it has a surface (top of the atmosphere) temperature of 59 K, which is similar to Neptune's. However, unlike Neptune the planet radiates as much heat from its surface as it receives from the Sun. This lack of a strong heat source is a surprise, as both Jupiter and Saturn radiate twice as much energy as they receive from the Sun and it is not fully understood how Uranus lost its internal heat source. Internal heat is a remnant from the birth of the planets; there are a number of reasons why it may have a different evolutionary path to Neptune. It is possible that the collision that Uranus underwent is the reason it no longer emits as much heat as Neptune, or maybe the stratification in the interior does not allow for any heat to escape to the surface or releases it at a different rate. Another perhaps less likely suggestion is that Uranus is an older planet than Neptune and has already emitted all its interior heat, therefore making it cooler today. Finally, the difference may be down to the rate of release. What can be deduced, though, is that two planets with remarkably similar origins can undergo different evolutionary paths. It is the differences between Uranus and Neptune that are just as significant to scientists, particularly when they consider the formation of the solar system. It was difficult to obtain observations of the precise shape and dimensions of the planet and early models of its structure were poorly determined. Even now there are different models of the interior of Uranus, and the mass of all the ice contained in the planet is still not known.

(Re-)Discovery of Rings

The discovery of the rings would come nearly two hundred years after that of the planet, not being found until the middle of the twentieth century. The path of Uranus through the solar system was now well known, as was the position of the background stars it would occlude (pass in front of). These events are of great value, as they offer information about a planet's position, diameter, shape and atmosphere. Gordon Taylor, an astronomer from the Royal Greenwich Observatory, was particularly keen to get an accurate diameter of the planet. This had proved tricky due to the fuzzy appearance at the edge of the disc, which held the potential for large errors. Taylor suggested that a future occultation would solve this problem.

Lunar occultations are frequent, but planetary ones less so. Taylor had recognized the value of an occlusion of Uranus in the early 1950s but the first candidate star that astronomers could use did not come into position until 10 March 1977, because Uranus was moving in an area of the sky that had very few suitable stars.[25] When an occultation occurs and the star appears to move behind the planet, the star will start to dim and it is this reduction in light that is key to astronomers' measurements. When watching the light curve, a slow dimming of the star means the planet may have an atmosphere; a quick dimming means it does not. If a planet has a ring system, the star will dim before it has even got to the disc edge and brighten again as it moves between the rings and the disc, before dimming yet again.

In January 1977, astronomers had the opportunity to measure the star SAO 158687 in the constellation of Libra as it passed within 0.8′ of the planet. This would give them a chance to hone their skills before the main event in March. Along with ground-based observations, the Gerard P. Kuiper Airborne Observatory (KAO) would be tracking the event in the sky using its onboard 0.9-metre infrared telescope.

The initial observations were concerning, as they showed that the planet was tracking in a different place to where it should have been. The light curve was not as they expected. Not to be deterred, on the night of 10 March 1977, the observing teams set

The Gerard P. Kuiper Airborne Observatory in flight.

up at locations across the world, from Western Australia, across the Indian Ocean and all the way to central Africa. Alongside these earthbound observers, the KAO was launched. The occultation of star SAO 158687 began and astronomers expected the light to drop as the star moved behind the planet, and then rise again as it emerged from behind. Astonishingly the brightness in the star winked, not once but five times, before it moved behind the planet. A number of satellites were considered and then discounted as causing this effect. There was only one possible explanation: Uranus had a multiple-ring system. A second occultation, in the following year, would enable the 200-inch Hale reflector on Palomar Mountain, California, to confirm these findings. The detection of the ring system was perhaps the most important revelation about Uranus between its discovery in 1781 and the Voyager flyby in 1986.

Now it is time to turn our attention to the compact spacecraft that revolutionized our understanding of the solar system and particularly of the ice giants, which, due to their incredible distance from Earth, had been difficult to scientifically understand.

TWO

Voyager 2 Flypast of Uranus

Voyager's story begins in 1965. While working at the Jet Propulsion Laboratory in California, the American aerospace engineer Gary A. Flandro was given the task of designing a method of exploring the outer planets by spacecraft. Flandro used the most up-to-date electronic computers to calculate the celestial mechanics involved, in contrast to the earlier laborious pencil and paper calculations. While working through the calculations for the outer planets, Flandro discovered that a rare alignment of Jupiter, Saturn, Uranus and Neptune would occur in the late 1970s. This event would not be repeated until the 2150s. It gave NASA the opportunity of sling-shooting a spacecraft via gravity assists from one planet to the next, a method which changes the velocity of the spacecraft by exploiting a planet's gravity. This would not only shorten the amount of time it would take to reach the outer planets, but reduce the costs involved.

Flandro had discovered the idea of what became known as the 'Grand Tour' of solar system exploration. While this plan was exhilarating to planetary scientists, NASA administrators resisted it and instead preferred a Jupiter-focused mission. Even though this was the 1960s, when NASA was experiencing an unprecedented – and, so far, never-to-be repeated – era of 'budgetless financing' because of the urgency of beating the Russians in the Cold War Space Race, support for spaceflight was largely limited to President

Engineers working on Voyager 2, 23 March 1977.

John F. Kennedy's goal of landing a man on the Moon. After the first Moon landing of the *Apollo* missions in 1969, the public's interest faded rapidly, and financial pressures began to loom. The Vietnam War and economic recession started to overtake the space programme as national priorities. By the time the last *Apollo* landing took place, in December 1972, President Richard Nixon declared,

'We will not return to the Moon.' NASA was desperately casting around for a new, exciting mission to justify its continued existence. The possibilities included a manned mission to Mars, a space station in Earth's orbit, and a space shuttle to transport people and materials to this space station. The preferred option came from an ambitious planetary mission, which would visit not just one planet but numerous ones, taking a tour through the solar system. The mission would become known as Voyager.

Funding had originally been sought for two pairs of spacecraft, which would take advantage of the rare 175-year planetary alignment and visit three planets each. One pair, to be launched in 1977, would fly to Jupiter and then swing past Saturn and on to Pluto. The other pair would set out in 1979 for Jupiter, Uranus, Neptune and Pluto.[1] The project was called the 'Grand Tour' because the timing was perfect. An individual trip to Neptune would take thirty years, whereas the opportunity offered by the Grand Tour would allow the spacecraft to reach the outer planets in less than thirteen, depending on the chosen trajectory and the gravity assists employed along the route. This Grand Tour would be saving NASA money, as the two pairs of spacecraft would be far more economical than numerous individual craft. Unfortunately, the overall plan was still deemed too expensive, and the proposal was scaled back to two spacecraft, Voyagers 1 and 2, which were to set out during the 1977 launch window.

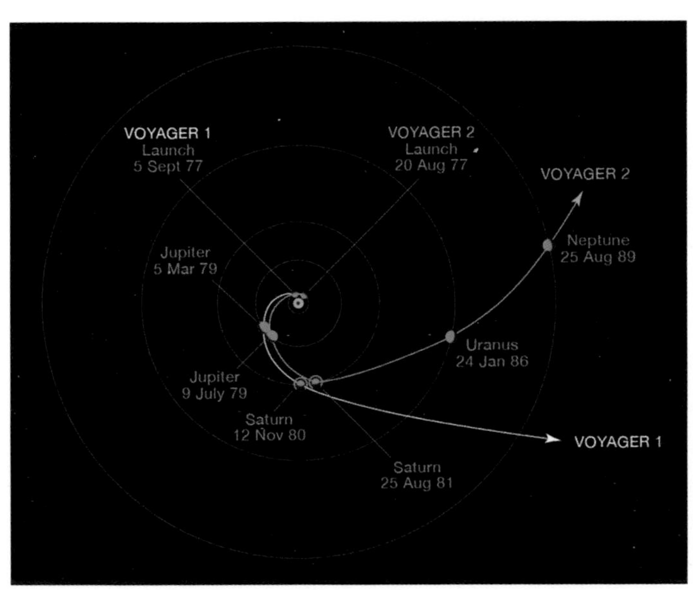

Voyager 1 and Voyager 2 trajectories to the outer solar system. Though both spacecraft encountered the gas giants Jupiter and Saturn, only Voyager 2 made flybys of the ice giants Uranus and Neptune.

The engineers at NASA's Jet Propulsion Laboratory

(JPL) started work on the design of the craft in January 1972, even though the final approval did not come through until May 1972. This approval arrived just before the recession had begun to bite, and thereby narrowly escaped cancellation by President Nixon's administration. Agreed funding came in for a smaller project, no longer a 'Grand Tour' but a Mariner mission to Jupiter and Saturn. However, the scientists had other plans and fully intended to take advantage of the rare planetary alignment. They anticipated that with careful budgeting and piecemeal additions, they could extend the mission to Uranus and Neptune. 'We understood at the time the enormous potential of this mission – that it could very well be one of the truly outstanding if not the most outstanding mission in the whole planetary exploration program,' explained Bradford Smith, leader of the Voyager Imaging Team.[2]

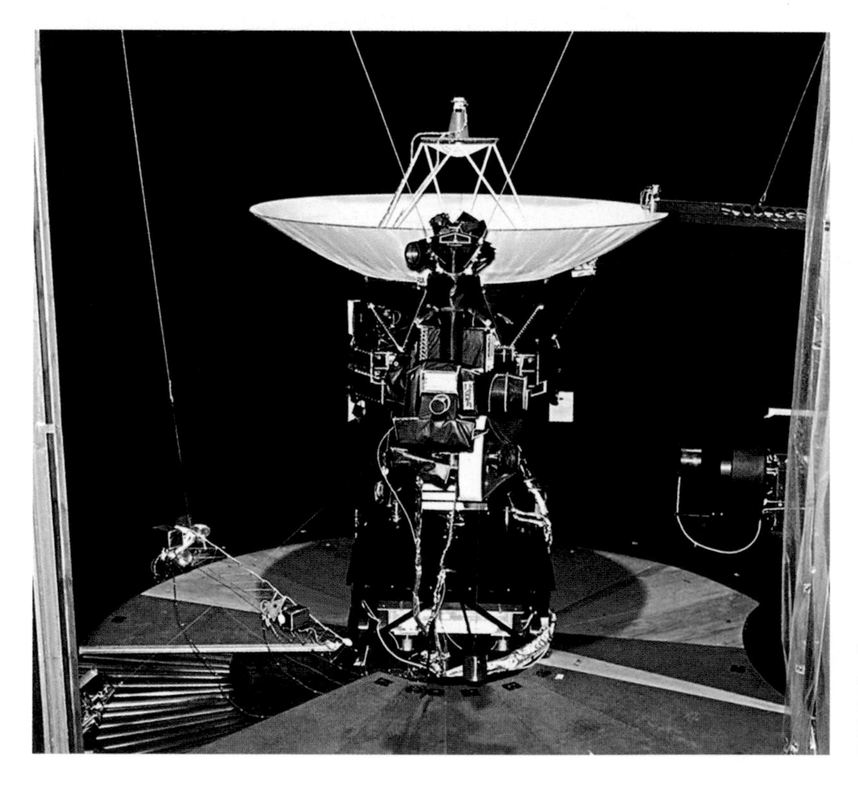

Voyager 2 test model, with all the hardware connected in a test chamber, at NASA's Jet Propulsion Laboratory, Pasadena, California.

Pluto would have to wait and eventually it received its own flyby by the spacecraft New Horizons in July 2015. The Voyager 1 itinerary had originally included Jupiter, Saturn and Pluto, but since Saturn's ring system and its large moon, Titan, were higher priorities than Pluto, in order to make a close flyby of Titan, Pluto was removed from the list. It was never on the Voyager 2 itinerary, which included flybys of Jupiter, Saturn, Uranus and Neptune. In theory, it could have a gravitational slingshot to journey on to Pluto from Neptune, but in order for that to be accomplished the spacecraft would have had to pass dangerously close to Neptune – indeed, inside the orbit of Triton. In 1973 Carl Sagan said,

> In all the history of mankind, there will be only one generation that will be the first to explore the solar system, one generation for which, in childhood, the planets are distant and indistinct disks moving through the night sky, and for which, in old age, the planets are places, diverse new worlds in the course of exploration.[3]

By the time Voyager 2 made its flyby of Neptune in August 1989, Sagan's own generation had begun to realize this dream.

In order to accomplish this mission, planners needed highly accurate predictions for the positions of each of the giant planets on the itinerary – Jupiter and Saturn for Voyager 1, and Jupiter, Saturn, Uranus and Neptune for Voyager 2. The challenge of providing this data fell to Erland Myles Standish Jr and his colleagues at the JPL, who specialized in dynamics and celestial mechanics. After discarding all pre-1910 positions for Uranus and Neptune, they produced new ephemerides (calculated positions over time) for Uranus and Neptune, which provided accurate positions for the forthcoming encounters. In addition, after the Voyager 2 flyby of Neptune, they used the spacecraft's newly determined mass of the planet, which had been revised downward by 0.5 per cent

– an amount comparable to the mass of Mars – to recalculate the gravitational effect on Uranus, and this reduced the residuals in orbital position to zero.[4] The theoretical and observed were in precise agreement. The spacecraft could now be aimed at their targets to a high degree of accuracy. After its flybys of Jupiter and Saturn, Voyager 2 would attempt a remarkably close, hazardous approach to Uranus, in which it would withstand exposure to the planet's magnetic field, pass close to the inner moon, Miranda, and then swing behind the planet and Uranus's slender rings before being hurled, by gravity assist, to Neptune.

Voyager 2 was launched from Cape Canaveral, Florida, on 20 August 1977. By 1979 it had reached Jupiter, then Saturn in 1981. It would be nine years into its mission before it reached Uranus in January 1986, before travelling towards and making its closest approach to Neptune on 25 August 1989. Moving through space at an average velocity of 19 km/sec (42,500 mi./hr), any deviation from the computed course would mean the planets would be missed. Not everything had gone according to plan on the mission. By the time it was approaching Uranus, the spacecraft was already eight years old. The first issue had been the failure of one of the receivers. This in itself was not a problem, because there was a backup receiver onboard. When the computer command system's (CCS) onboard computer switched over to the standby receiver, it became obvious to the team that this was not working either, due to a faulty capacitor. The craft had been pre-programmed to compensate for a Doppler shift in the signals sent from Earth. As it travelled through the solar system, the amount changed and, in turn, this caused problems with the radio receivers. The issue with the receivers would be solved by engineers tuning into a short-wave radio band at an exact frequency to communicate to the craft.[5] Near the closest approach to Saturn the azimuth motion of the scan platform, which carried all the cameras, became jammed and was impossible for the team to fix, thus losing key images from the planet and its satellites.

It would take two days for them to solve the issue, and even then the scan platform would not be working optimally. The team were able to find ways to work around the problem. They learnt how to orientate the craft as well as move the platform, allowing them to aim the cameras in directions which the jammed scan platform had otherwise made impossible.

Imaging Uranus would present further challenges. The pictures would have to be long exposures, as Uranus receives only one four-hundredth of the light that we do here on Earth. Capturing this light would require the cameras to keep their shutter open for as long as 96 seconds. There was the danger of images being blurred or smeared due to the rapid motion of the craft. The team employed various techniques to overcome this and, in particular, instructions to the onboard attitude control system were altered, with the intention of being able to track targets with a high degree of accuracy. By the time the craft reached Uranus, it would be 3 billion km away from Earth and transmission would take 2 hours 45 minutes, so it would be virtually impossible to carry out repairs and send new instructions. Everything had to be in order before the flypast.

More Rings

Voyager 2's approach to the planet allowed an opportunity to study the ring system in more detail. Rings around Uranus had been identified, initially by James L. Elliot, Edward W. Dunham and Jessica Mink during the occultation of the star SAO 158687 in 1977. At least fifteen occultation events were then observed, revealing nine narrow rings encircling the planet. They were designated epsilon (ε), delta (δ), gamma (γ), eta (η), beta (β), alpha (α), 4, 5 and 6, with '6' being the closest to the planet and ε being the furthest.

The Voyager 2 flypast was to add to this tally with another three rings, 1986U1R, 1986U7 and 1986U2R, bringing the total to twelve.

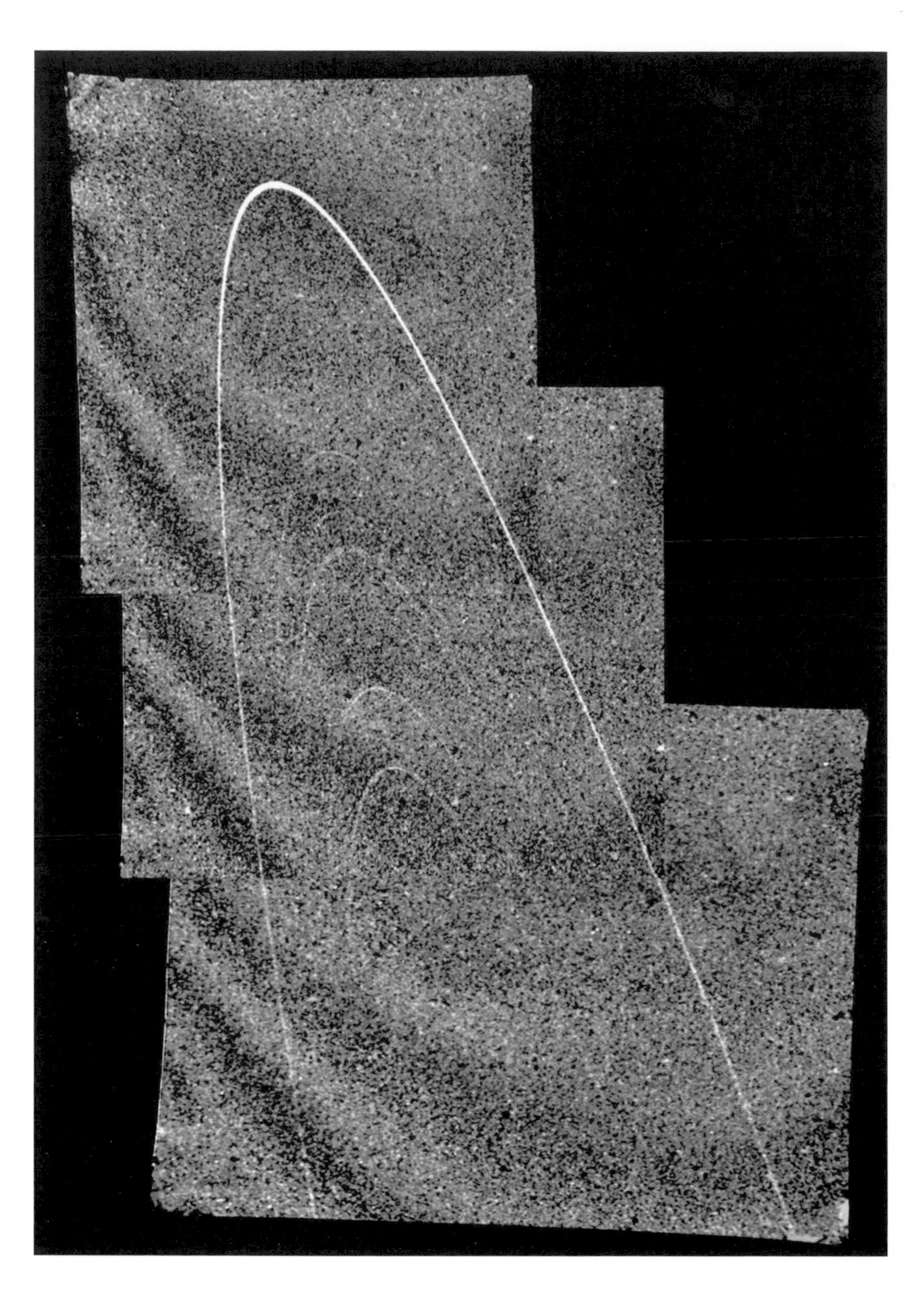

The ring system imaged as Voyager 2 approached the planet Uranus in 1986.

It is now known that Uranus has thirteen rings, with the last being discovered by the Hubble Space Telescope, and there is a possible fourteenth ring which does not completely encircle the whole planet. A table of their properties can be seen in Appendix III.

Thanks to the axial tilt of the planet, the rings can be viewed face on, but this depends on the position of the planet and Earth in relation to its orbit around the Sun. The rings differ in composition to the rings of Saturn, which are made of water ice. Instead, Uranus's rings are dark and this is most likely due to carbon compounds. The brightest ring, Epsilon, is composed of boulders that can reach 1 m in diameter, although the ring itself has a thickness of less than 150 m.[6] This is typical of the rings, as they are all narrow and thin. It is thought that the ring system is relatively young – about 600 million years old. It probably formed from a number of satellites that broke up during the collision that knocked Uranus on its side.

In 1983 Carolyn Porco joined the faculty of the Department of Planetary Sciences at the University of Arizona and became a member of the Voyager imaging team. She is now well known for her work on the joint NASA/ESA Cassini-Huygens mission to Saturn. At that time she was helping to calculate the camera exposure times that brought us our only up-close looks at the rings of Uranus and, later, Neptune. Her description of the behaviour of ringlets around Saturn would equally show how the moons Cordelia and Ophelia act as shepherd moons, creating the mechanism that would keep the two outer rings of Uranus from dispersing. The Epsilon ring has sharp edges which are defined by the moon Ophelia on the outside and Cordelia on the inner edge.[7] Ring arcs have formed along the outer ring. These arcs are areas of greater or enhanced material which form along a ring. Four have been identified in the Voyager data but two of these have since disappeared, although it is probable they will be replaced in the future.

Magnetosphere

Five days before Voyager 2 reached its closest approach of the planet, the team obtained the first direct evidence of a magnetosphere surrounding the planet. Radio emissions detected the magnetic field at 27.5R Uranus (where R Uranus = the radius of Uranus) and Voyager then subsequently crossed the bow shock at 23.5R Uranus. On reaching 18R Uranus, the craft was to finally enter into the magnetosphere. The magnetosphere forms a region around the planet where charged particles are affected by the magnetic field of the planet. These energetic particles are trapped within the field lines, which act as a magnetic dipole. The magnetic field of a planet forms a windsock region, of which the head faces the Sun and the tail streams away from the Sun behind the planet. The field is shaped and buffeted by the solar wind. Both Jupiter and Saturn have strong magnetic fields and it seemed obvious that Uranus would follow their example.

However, when it entered the magnetic field, Voyager 2 discovered a magnetosphere that was unlike the other gas giants. Ed Stone, the Voyager lead project scientist, said,

> We knew Uranus would be different because it's tipped on its side, and we expected surprises. Then we got to Uranus and saw that the [magnetic] poles were closer to the equator. Neptune turned out to be similar. The magnetic field was not quite centered with the center of the planet.[8]

This was due to its position in relation to the geographic poles. It was at a 59° angle compared to its spin axis, which at that time suggested a source slightly different to the other known magnetospheres. This inclination is the most of any magnetic field in the solar system and, at times, the magnetosphere opens, where the field lines are attached to the planet at only one end; at other

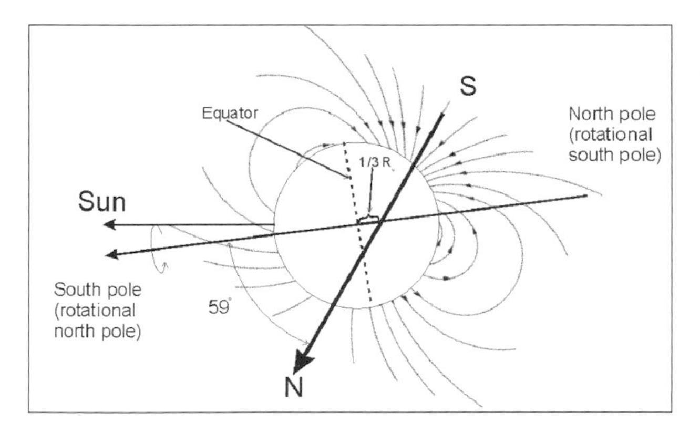

Voyager 2 observes the magnetic fields on Uranus and finds the location of magnetic N and S, which can be seen in this sketch.

times the magnetosphere is closed, where the field lines loop and both ends are attached to the planet. When it is in the open state, it leads to an unusual occurrence: the magnetic particles will be driven by the solar wind. The inclination also means that the field rolls asymmetrically, and this allowed the Voyager team to observe the planet's rotation. The resulting period of rotation was found to be 17.24 hours. A day on Uranus thus takes seventeen hours and fourteen minutes but its rotation is retrograde: the planet rotates clockwise, or east to west as seen from its north pole, in the same way that Venus does.

The reason why the magnetic field is in this position has been extensively debated. One suggestion is that the early collision the planet had, which knocked it on its side, also changed the position of the magnetic poles. An alternative suggestion is that Uranus's magnetic poles are undergoing a period of reversal. So the north and south poles of the magnetic field are in the process of flipping. This has happened here on Earth in the past, most recently between 41,560 and 41,050 years ago.[9] Due to the complexities of the magnetic field, auroras can be seen at a wide range of latitudes. Although they are difficult to observe given the distances involved and they were not observed by the craft, they have been imaged by the Hubble Space Telescope subsequently.

Atmosphere

An alarm sounded just four days before the encounter, and one of the onboard computers flagged up a memory failure. The team worked frantically trying to fix the problem and were finally able to resolve the issue. Scientists had started to image the planet from

November 1985 when the craft was still 103 million km away. In these early images Uranus did not appear to have any markings on its surface. It was a wonderful turquoise blue but seemed to lack any distinctive features. It is now known that this unexpected uniformity was the result of the planet being at the solstice of its southern summer. Voyager 2 flew past when Uranus's south pole was pointing directly at the Sun. Under permanent sunlight, the temperatures were uniform and methane would have created a high-altitude haze, masking any activity underneath. Uranus's surface is very cold; it is the coldest planet in the solar system and at the cloud tops the temperature is only 49 K. It was difficult to draw conclusions about the effect of seasons on Uranus on the basis of the evidence from the Voyager mission alone.

The climate on Uranus is calm compared to the other gas giants, although there is a huge amount of variation over the 84-year orbit. At the solstice, the polar regions appear bright, whereas at the equinoxes darker equatorial bands are visible and atmospheric temperatures have their greatest range. It was just three days before the craft had its closest encounter when banding was noticed on the surface of the planet in the equatorial area. A bright southern cap and darker equatorial bands were imaged. It unfortunately did not observe the north pole, which was pointed away from the Sun and thus in darkness; this has only been examined subsequently by the Hubble and Keck telescopes. Voyager 2 eventually imaged clouds within the atmosphere of Uranus on Tuesday 23 January 1986. Ed Stone spoke at the daily news briefing: 'This is the first time discrete clouds have ever been detected in the atmosphere of Uranus.' Noting that the clouds at different latitudes were moving, he concluded that 'there are winds there.'[10] This was the first evidence for winds in the atmosphere of the planet. The clouds were imaged by Voyager at latitudes between 20 and 45°, where sunlight penetrates to greater depths through the haze, and formation of clouds could be seen. The atmosphere on Uranus mostly comprises

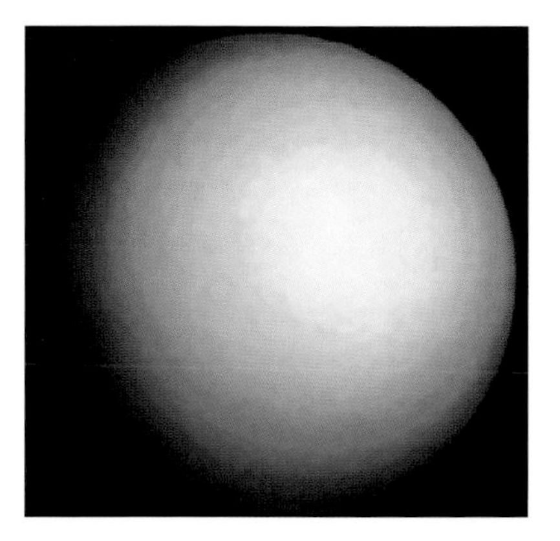

A departing image of Uranus taken in January 1986, showing the photochemical smog in the upper atmosphere, which results in the greenish colour of the planet.

hydrogen and helium; this is enriched by other volatiles such as methane at lower levels. The clouds are found in a region called the troposphere, which extends up to 50 km above the surface of the planet and is characterized by a reduction in temperature with an increase in height. It has a pressure of 100 to 0.1 bar; it was at 1 bar that the methane clouds were found by Voyager 2. The flypast observed clouds moving in an east–west direction rather than north–south; this is in keeping with Saturn and Jupiter.[11] It was soon calculated that the clouds were moving at speeds of up to 480 km/hr; the speed the clouds rotated around the planet varied between 15 hours and a slightly slower 16.25 hours.

In the stratosphere of Uranus the temperature increases with altitude and this increase in temperature allows hydrocarbons such as ethane to form through interactions with methane clouds, which contain methane ice crystals. The temperature rises from a very cold 53 K to 850 K where it meets the thermosphere. This layer of the atmosphere ranges from 50 to 4,000 km. In addition to methane, there is carbon monoxide, carbon dioxide and water vapour in the stratosphere. When observing the planet through methane wavelengths, a high layer of haze can be seen along with the clouds. This layer of haze is made from acetylene and ethane particles. It is the absorption of methane at these high altitudes that gives Uranus its greenish colour.

The haze layers are much brighter on the sunlit southern pole of Uranus, as taken by Voyager 2, published 8 May 1999.

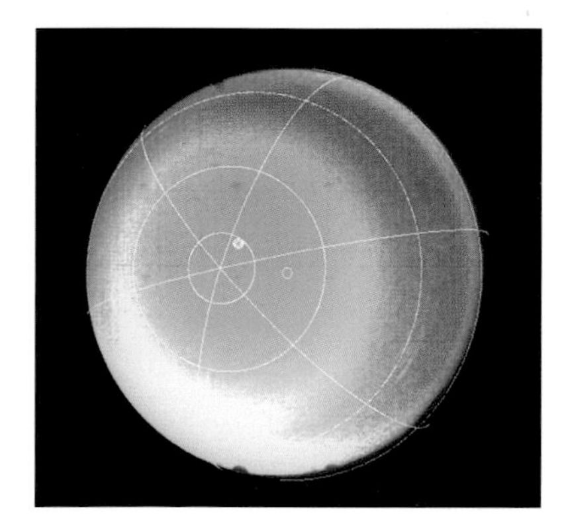

Observing the clouds in different wavelengths has allowed scientists to delve deeper into the weather systems on Uranus. One unexpected finding from the flypast was the amount of ultraviolet light emitted. The source of this ultraviolet radiation was the upper atmosphere of the planet, its thermosphere. This is also known as dayglow or electroglow. The reason for this emission was a mystery at the time of Voyager 2.

It can only be seen on the sunlit southern half of the planet and the emissions are in the far ultraviolet range of 90–140 nm. They extend across a huge area of the planet, over 50,000 km^2. It is now thought this is due to molecular hydrogen being excited by sunlight. The thermosphere has an unusually high temperature of 800–850 K, which is also poorly understood. Auroral activity and incoming particles from the magnetosphere raise temperatures, and the aurorae were thought to be important in contributing to the raised temperature of the upper atmosphere of Uranus.

The ionosphere is the final layer of the atmosphere. When Voyager 2 was occultated by the planet Uranus, by moving behind it when viewed from Earth, radio waves in the 13-centimetre and 3.5-centimetre bands were able to measure the height and density of the ionosphere. The radio occultation experiment showed the ionosphere to extend out to 10,000 km.[12] Solar ultraviolet radiation sustains the ionosphere, as it ionizes the molecules and atoms, creating free electrons.

Structure of Uranus

The closest approach of Voyager 2 would occur on 24 January at 17 hours 58 minutes 51 seconds UT (Universal Time). When it arrived at the planet, the interior of Uranus was little understood. Measurements of the radius, mass and gravitational field by Voyager were all used to gain an understanding of the interior of the planet, and during the flypast all these elements were measured in great detail, alongside the shape of the planet itself. The simplest models suggest Uranus has three main layers: a solid core, an ice mantle and a gaseous atmosphere. In the centre is a small rocky core made from silicates/iron-nickel; this has a radius of less than 20 per cent of the total. It is surrounded by the mantle, which consists of ices made from water, ammonia and methane. This layer makes up most of the planet and contains about 13.4 Earth masses of material. The final 20

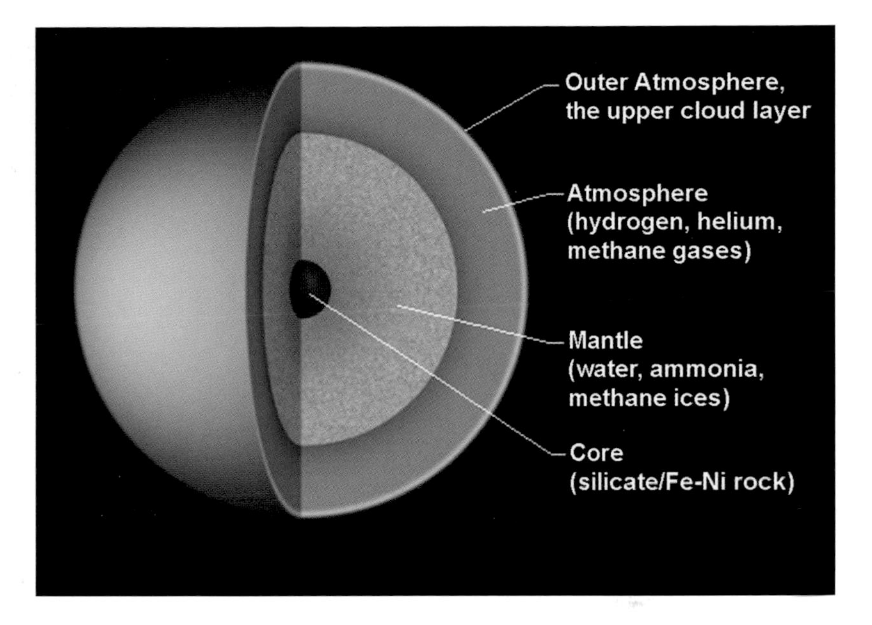

Structure of Uranus from outer atmosphere to the planet's core.

per cent of the radius is the atmosphere and it contains just 0.5 Earth masses; it is made of hydrogen, helium and methane gas.

There is still much to be discovered and alternative models suggest there could be a blurring of layers, with hydrogen from the atmosphere and rocky material from the core mixing through the ice layer. Molecules of methane are subject to high pressures within the interior; as they reach within the lower layers, where the pressures and temperatures increase, they may break down, with the carbon part of the molecule raining down and creating an ocean of diamonds at the bottom of the mantle. Even today questions remain about the boundary layers and composition of the interior.

New Moons

During the Voyager 2 flypast a number of new satellites were discovered. Only five had been known before the mission and a further ten were observed. All of these new moons were found to be small in size, with Puck being the largest at a diameter of just 162 km. Alongside Puck came Juliet, Cordelia, Ophelia, Bianca, Desdemona, Portia, Rosalind and Cressida – all named after characters from plays by William Shakespeare. The satellite Belinda was named after the character in the poem *The Rape of the Lock* by Alexander Pope.

They all orbit between the planet and the already-known five larger moons (Miranda, Ariel, Umbriel, Titania and Oberon). The images returned by the Voyager cameras would be spectacular. The craft passed the moon Miranda at a relatively close 28,260 km and returned images of a unique and interesting world.

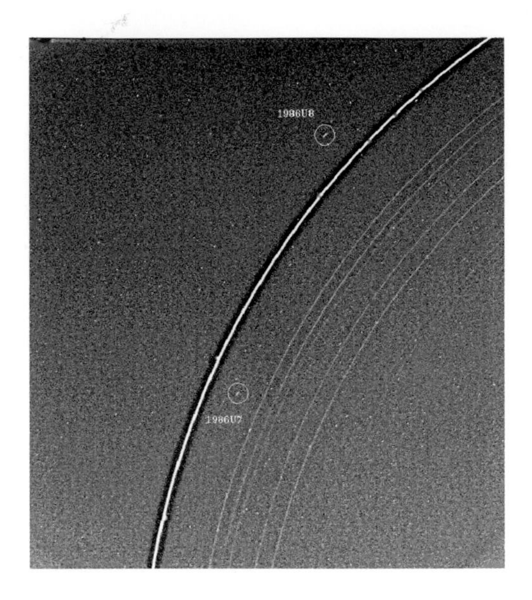

Two of the new satellite moons that were discovered by Voyager: 1986U7 (Cordelia) and 1986U8 (Ophelia).

The Voyager spacecraft took more photographs than any previous mission. The importance of having cameras on board had been identified in the design stage by Harris M. 'Bud' Schurmeier of the JPL, who had 'understood both the non-scientific and the scientific importance of imaging the planets and their satellites'.[13] The images of the moons, considered in greater detail in Chapter Four, brought a whole new set of worlds into focus for the science community to study and the public to look at, capturing everyone's imagination with a vibrancy and clarity yet to be repeated.

Departure

On 25 January 1986, the spacecraft turned to take a parting shot of the planet. It would result in an image of a beautiful crescent of Uranus shining in sunlight, with the rest of the planet in shadow. The craft would have to make a course correction on 14 February to set it on a direct path to its final destination, Neptune. The accuracy needed was almost inconceivable, and the calculations worked out brilliantly. As science writer Mark Littman said, when going from Uranus to Neptune, Voyager 2 would have to 'hit a point in space at a point in time with phenomenal accuracy. To miss "corridor centre" at Uranus by a mile would create an error of 6,437 kilometres at Neptune . . . The accuracy required . . . was a feat comparable to a golfer sinking a thousand-mile putt.'[14]

Voyager 2's data from the flypast of Uranus is still being used by scientists to delve deeper into aspects of the planet. In 2014 Erich Karkoschka, a planetary scientist at the University of Arizona, stacked 1,600 images taken during the flypast, discovering dozens of missed cloud features. He revealed that the southern hemisphere moved differently to the rest of the planet, rotating by as much as 15 per cent faster than features on the northern hemisphere.

> The unusual rotation of high southern latitudes of Uranus is probably due to an unusual feature in the interior of Uranus. While the nature of the feature and its interaction with the atmosphere are not yet known, the fact that I found this unusual rotation offers new possibilities to learn about the interior of a giant planet.[15]

Three decades after the flypast in 2019, space physicists Gina DiBraccio and Dan Gershman at NASA's Goddard Space Flight Center re-examined the readings taken by Voyager as it travelled through the magnetic field of the planet. They discovered a jagged zigzag in the strength of the magnetometer readings and realized the spacecraft had encountered a plasmoid. This is a giant magnetic bubble which would be responsible for the atmosphere escaping into outer space; it also causes the atmosphere to lose mass. This plasmoid is likely to have been one of many. As DiBraccio explains, this could be the main way Uranus loses mass over time, with up to 55 per cent of atmospheric loss happening in this way. Many questions still remain about observations made by Voyager 2; the

Parting shot showing Uranus as a crescent, taken on 25 January 1986.

effect on Uranus from the escape of plasmoids over time is just one of them. 'Imagine if one spacecraft just flew through this room and tried to characterize the entire Earth,' says DiBraccio. 'Obviously it's not going to show you anything about what the Sahara or Antarctica is like.'[16] We will have to wait for another spacecraft to visit the planet to gain the answers to some of these questions, although the work of Earth-orbiting and ground-based telescopes, considered in the next chapter, have brought substantial advances in our understanding of Uranus.

THREE

Uranus after Voyager

The Voyager 2 flypast of Uranus provided scientists with an incredible amount of data, which radically improved our understanding of the planet. Once the spacecraft had continued on its journey towards Neptune it became the turn of the ground-based and space satellites to study the distant ice giant. Viewing Uranus from Earth had always been a challenge, in part due to our atmosphere being opaque at some wavelengths and in part owing to the incredible distance involved. However, improvements in ground-based technology and telescopes such as the Keck, ALMA (Atacama Large Millimeter/submillimeter Array) and Subaru, alongside the launch of the space-based Hubble Space Telescope and Chandra X-Ray Telescope, allowed scientists to address some of these challenges. This in turn increased our understanding of Uranus in the absence of a new space probe.

New Moons and Rings

Since it was launched in 1990 the Hubble Space Telescope (HST) has revolutionized our view of the universe. Imaging Uranus between 2003 and 2005, HST discovered two new rings: R/2003 U1 and R/2003 U2; these were subsequently named the Mu (μ) and Nu (ν) rings. Mark Showalter of the SETI Institute described how 'the new discoveries dramatically demonstrate that Uranus has a youthful and

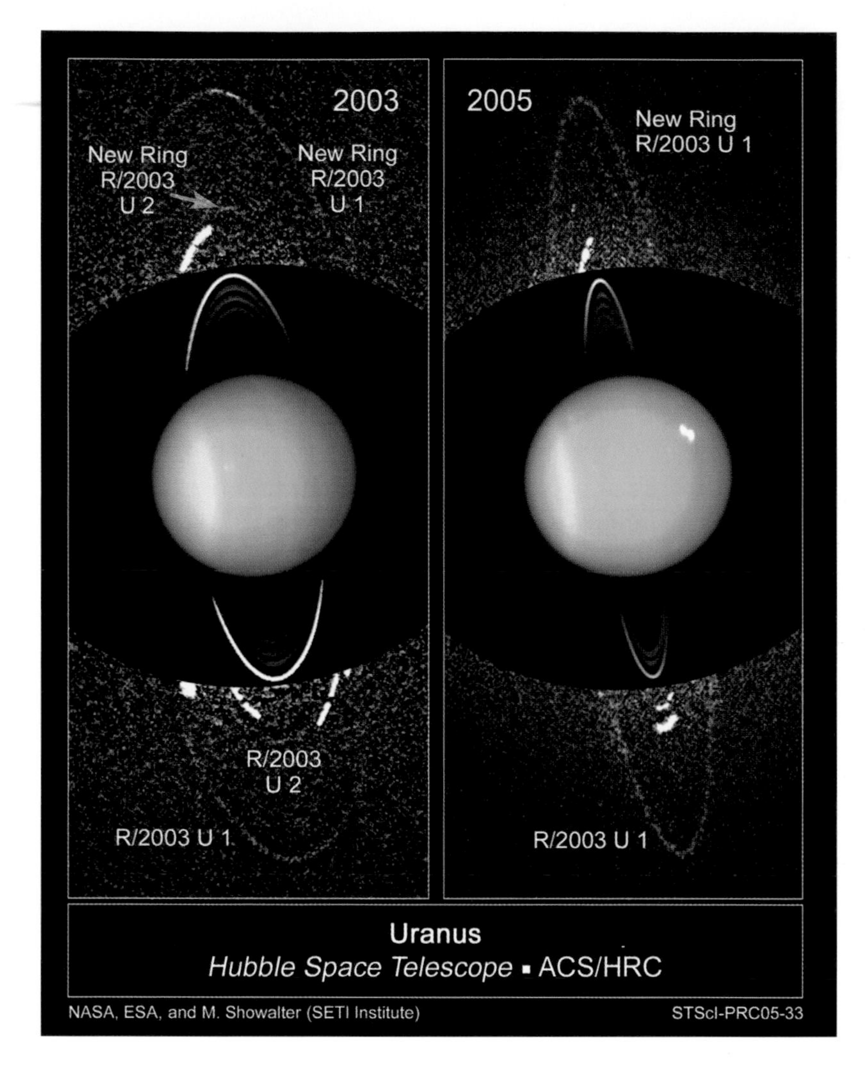

Discovery of the new ring systems R/2003 U1 and R/2003 U2 by the Hubble Space Telescope.

dynamic system of rings and moons, until then nobody had a clue the rings were there, we had no right to expect them.'[1]

The outer ring, Mu, was accompanied by a small moon, S/2003 U1, which was discovered by HST in 2003. A second moon, S/2003 U2, was observed at the same time and these moons were subsequently named Mab and Cupid. They are respectively 16 km and 8 km in diameter and were missed by Voyager 2's cameras.

It is remarkable how HST can spot these tiny objects in an area dense with other moons. Mab is a possible source of the dust that made the ring and continues to replenish it. The instability in the dynamics of the moon and ring arrangement mean that it is rapidly changing and very young, providing insight into the subsequent evolution of the entire system. These moons will be discussed in further detail in the next chapter.

Ground-based observations of the new rings were made by the Keck Telescope during 2005. Using the infrared region of the electromagnetic spectrum, Imke de Pater of the University of California, Berkeley, Heidi B. Hammel of the Space Science Institute in Boulder, Colorado, and Seran Gibbard of Lawrence Livermore National Laboratory were able to observe the faint rings around the planet. Looking at the two new rings, Imke de Pater noted how the inner ring glowed red and the outer glowed blue, saying:

> We have been puzzled by the fact that we cannot detect the outer ring, even though it is more than twice as bright as the inner ring in the Hubble data. The difference is probably caused by the fact that we are working in the infrared, at a wavelength four times longer than that used by the Hubble telescope in the rings' discoveries. This seems to suggest a color difference between the two rings, which would indicate that their constituent particles are very different.[2]

In 2007 Uranus approached its equinox, when the Sun would shine directly above Uranus's equator. The rings would appear edge on and allowed for a rare opportunity to observe the outer dust rings via transmitted light and more detailed studies to be undertaken of the ring system as a whole. One mystery that remained was the faint heat signature the rings emitted compared to the surrounding space; the reason behind this would not be solved until much later, by a set of ground-based observations made in 2017.

The Temperature of the Rings

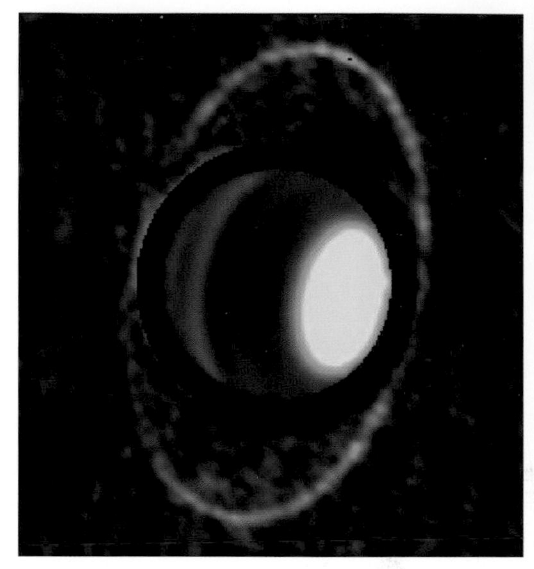

The thermal emission from the rings of Uranus, taken with Atacama Large Millimeter/submillimeter Array (ALMA) and the Very Large Telescope (VLT).

By examining planetary features at different wavelengths, astronomers can obtain more information about the object being observed. This is very much the case with Uranus. In 2017 ALMA and the Very Large Telescope (VLT) in northern Chile turned their attention to Uranus and imaged the rings in the infrared. Looking at the thermal emissions or heat signature, they determined a ring temperature of 77.3 ± 1.8 K, about 10° warmer than the planet itself.[3] The heat emitted by the rings was unexpected, as it is the only ring system in the solar system where this phenomenon occurs. De Pater, at the University of California, deduced that Uranus's rings had to have a different consistency from Saturn's rings. The rings lacked dust-sized particles, allowing them to reflect light to a much greater extent than the rings around other planets. De Pater said:

> Saturn's mainly icy rings are broad, bright and have a range of particle sizes, from micron-sized dust in the innermost D ring, to tens of meters in size in the main rings . . . The small end is missing in the main rings of Uranus; the brightest ring,Epsilon, is composed of golf ball-sized and larger rocks.[4]

Decades of Storms

Uranus's long seasons drive its storms and one particular storm has been raging in its atmosphere for decades. Called the polar hood, this is a large region extending over the pole, which shines white in direct sunlight. Polar hoods are composed of methane ice crystals and may develop from atmospheric flows between this region and

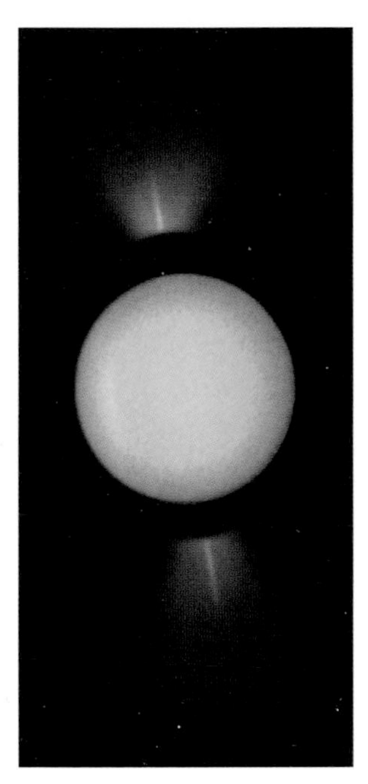

Uranus at equinox in 2007, where sunlight bathes the northern and southern latitude equally. The southern half is to the left of the image, with a band visible.

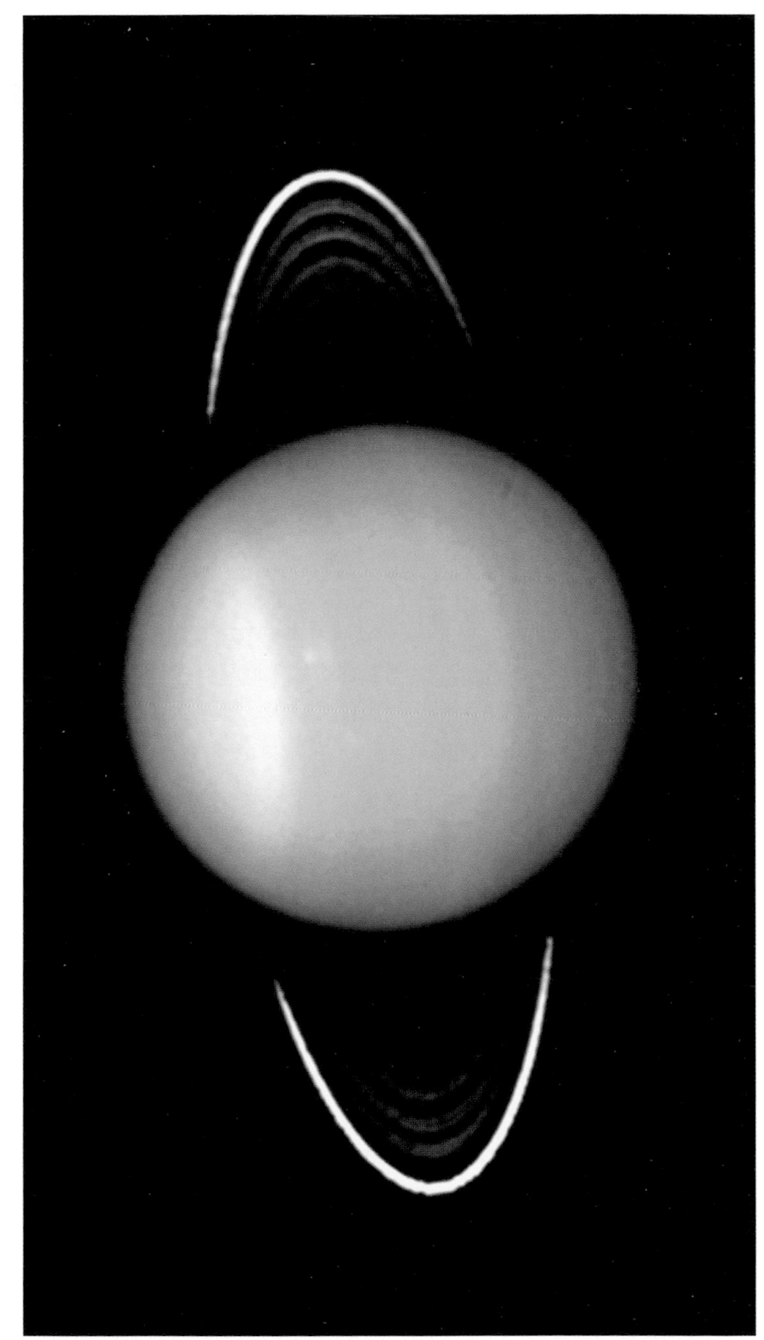

The white band on Uranus as imaged by the Hubble Space Telescope in 2003.

the neighbouring one.[5] At the 2007 equinox, for the first time in 42 years, the northern and southern hemispheres of the planet received an equal amount of sunlight. Up to then, there had been an asymmetry to the clouds that had formed on the surface of the planet, due to its tilt and resulting long seasons. The bright band, which was visible up to and at the equinox, was repeatedly imaged over the south pole by HST. Astronomers wondered what would happen when the south pole turned away from the Sun. Would an equally strong polar hood form over the northern hemisphere, once it was facing the Sun? It would, and a northern hood would start to develop as the pole was bathed in light. Then, just before the equinox arrived, a dark spot was imaged on the surface.

A Dark Spot

A team led by Lawrence Sromovsky of the University of Wisconsin–Madison imaged a dark spot on Uranus, using HST, in 2006, which was two-thirds as large as the United States of America.[6] Although scientists had assumed that Uranus's lack of a strong heat source would produce an atmosphere that would not be as active as other gas giants, they were beginning to realize this was not the case. Images taken on 23 August 2006 by HST confirmed the appearance of a dark spot. It was at a latitude of 27° in Uranus's northern hemisphere. This region had been in darkness until the approach of equinox – as the atmosphere came out of the shadows and was bathed in seasonal sunlight, an increase in temperature may have driven the formation of the spot.

There had been earlier unconfirmed sightings of dark features. The first such observation was by Louis Thollon and Henri Joseph Anastase Perrotin in 1884 at the Nice Observatory. Then came a tantalizing image of a possible dark feature spotted in low-contrast ultraviolet images from Voyager 2. The dark feature showed up when a methane filter was used at 619 nm, but this did not show

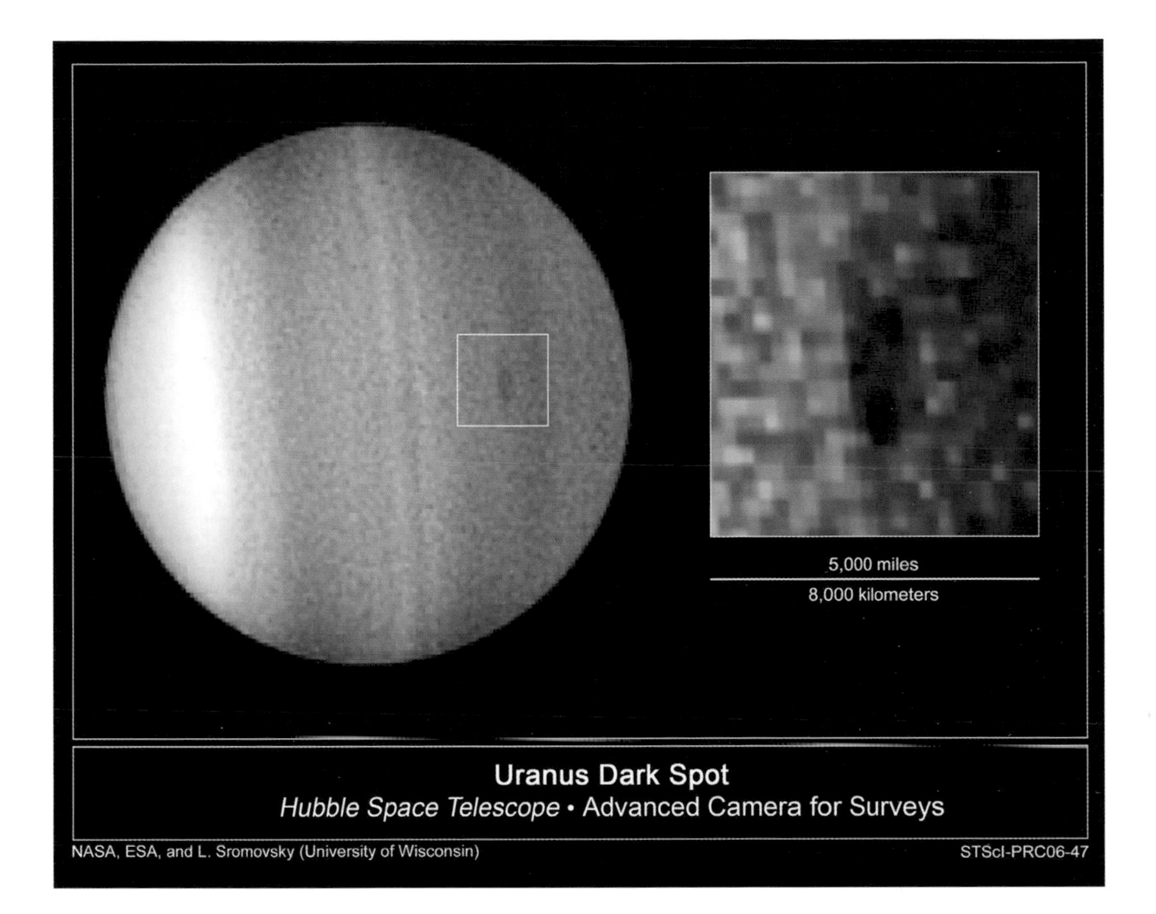

A dark cloud imaged on Uranus in 2006 by the Hubble Space Telescope.

in the wavelength band at 614 nm. At the time, it was suggested this was due to a thinning in the cloud layer below the haze.[7] A third possible dark spot was identified on a near-infrared image from the Multiple Mirror Telescope at the Fred Lawrence Whipple Observatory on 30 May 1993, but this observation was also uncorroborated.[8]

Given the absence of any previous sightings of dark features by HST in its annual reviews of Uranus, scientists concluded these spots were short-lived and had formed due to the season and climate conditions of the planet at that time. In much the same way we have seasons here on Earth, which change the temperature and cloud cover, the formation of dark spots could herald the onset of spring on Uranus, this time in the northern hemisphere. Planetary scientist Hammel, working at the Space Science Institute in Boulder, Colorado, said, 'We have hypothesized that Uranus might become more Neptune-like as it approached its equinox. The sudden appearance of this unusual dark feature suggests we might be right.'[9]

The axial tilt means that hemispheres of the planet are alternatively bathed in light and then plunged into darkness for many years at a time, leading to extreme seasonal variations. When sunlight shines on a previously dark part of the atmosphere, it responds to the changes in light, the temperature rises and spots can form. How long would the spot last? When HST renewed its gaze on Uranus the following year, the spot was no longer there. Tracking the weather patterns in this way over time gives astronomers an opportunity to discover trends in atmospheric dynamics.

The 'Berg'

Throughout the 1990s and 2000s a major cloud feature called the 'Berg' was imaged in the upper atmosphere of Uranus. It was called the Berg because it looked like a massive iceberg. It started as a group of clouds which were seen in the southern hemisphere near the edge of the polar hood at a southern latitude location of about 34°. This was a storm high in the atmosphere, with one main cloud body which was accompanied by a number of smaller companion clouds. Imaged over a number of years, it oscillated around a common central area of a few degrees; it may have been doing this for decades.[10] Then it went on a five-year migration northwards. In 2009 it was very faint and almost on the equator at a latitude of 5° south, and then it no longer appeared in photographs taken of the planet.[11] Before the Berg reached the equator, it had dissipated.

The 'Berg', a bright white feature on the southern latitude of Uranus, can be seen clearly in the right-side image, taken in 2004.

Subsequently, in 2014, the Keck Telescope imaged a number of huge storms in the upper atmosphere in the infrared. Over the course of two days, there was an evident increase in the size and number of bright spots in the atmosphere of Uranus. The rapid development of the storms showed that Uranus had a dynamic weather system and was not the staid world astronomers had considered it to be. The features were remarkably similar to the Berg, with de Pater remarking,

> We are always anxious to see that first image of the night of any planet or satellite, as we never know what it might have in store forus. This extremely bright feature we saw on 6 August 2014 reminds me of a similarly bright storm we saw on Uranus's southern hemisphere during the years leading up to and at equinox.[12]

Eventually, like the Berg, the storms may drift towards the central latitudes and then dissipate, although they were still in existence in 2020.

During its annual observations made of the planet in 2018, HST showed a remarkably well-defined north polar cloud cap in greater detail than ever before. The north pole was now in constant sunlight, and thanks to the extreme tilt of Uranus, the Sun will not set over this region during the long summer. The polar hood became more prominent than ever. In the following year, it became so bright that amateur astronomers were able to image the white region, along with another patch of cloud which is visually prominent. A light-coloured band extending around the planet just north of the equator was also visible in the 2018 images. Why this

The north pole cloud cap shown in more detail than ever before, as taken by the Hubble Space Telescope Wide Field Camera 3, November 2018.

band is so narrow in width remains a mystery, as both Uranus and Neptune have broad westward-flowing jet streams of wind. Once again, HST's annual observations of the planet showed that it was more dynamic than scientists had believed when the planet had earlier been imaged by Voyager 2.

Solar Storms and Aurora

In 2011 the HST imaged an auroral event on Uranus for the first time, a feat then repeated the following year. Using HST's Imaging Spectrograph ultraviolet camera, a team led by Laurent Lamy, at the Observatoire de Paris in Meudon, France, witnessed an intense event in November 2014, which had been set in motion

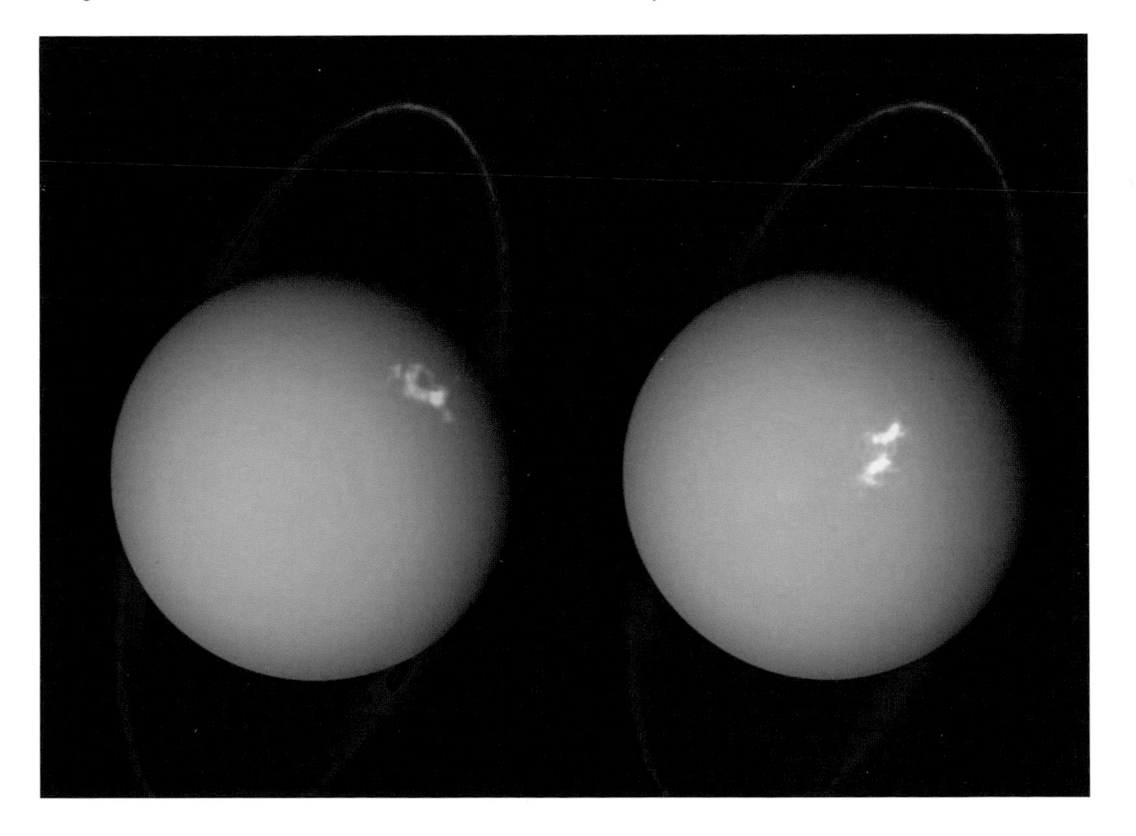

The aurora on Uranus imaged by the Hubble Space Telescope in composite with a Voyager 2 image.

by two powerful solar coronal mass ejections. These solar events sent charged particles across the solar system which, on reaching Uranus, were accelerated in the magnetosphere and driven towards the magnetic poles. As when the solar wind interacts with Earth's magnetosphere, a spectacular visual display around the poles was created. The location of the magnetic poles was unknown on Uranus since the last confirmation of their location by the Voyager 2 flypast, and effectively they had been lost. The position of this auroral display enabled astronomers to discover the current location of the magnetic poles. The aurora events did not last very long, only shining for a few minutes at a time, in contrast to the hours-long displays that we can enjoy here on Earth.

The Discovery of X-Rays on Uranus

In early 2021 NASA's Chandra X-Ray Observatory discovered Uranus emitting in the X-ray region of the electromagnetic spectrum. A team led by William Dunn at the UK's Mullard Space Science Laboratory were able to show from earlier Chandra data that X-ray flares had been emitted from the planet in August 2002 and November 2017. X-ray emissions have been observed from numerous planetary and cometary bodies, including Venus, Earth, Mars, Jupiter, several of Jupiter's moons, Saturn and Pluto. Previously the German Röntgensatellit (ROSAT) X-ray satellite telescope had suggested that there were no X-ray emissions. The heightened sensitivity of the Chandra X-Ray Observatory has shown otherwise. Dunn explains the importance of the finding, saying, 'The study of X-ray emissions from planets provide key and often unique insights into a variety of characteristics of the system. Most relevant for Uranus, these include: atmospheric, surface and planetary ring composition through fluorescence.'[13] The Mullard team have postulated that the emissions come from the Sun's light, which is interacting with Uranus and scatters the X-rays, in the same

Uranus glowing with X-rays.

way that it is scattered at Saturn and Jupiter. There is the suggestion that there may be a secondary emission of X-rays when charged particles, which surround the rings around Uranus, collide with material from the rings and start to glow in the X-ray part of the spectrum. Another possibility is that the charged particles that are interacting with the atmosphere during an aurora event may be the source. This would be similar to what happens here on Earth, where X-rays are emitted during aurora events.

The Tilt of Uranus

In 2006 the Canup–Ward model was able to demonstrate the position and density of the satellites found around Uranus, yet the model was unable to show the satellites' configuration once the 98° tilt was taken into account.[14] This was unexplained until 2020, when a team of astronomers led by Shigeru Ida, a professor at the Earth-Life Science Institute (ELSI) at Tokyo Institute of Technology in Japan, tried to discover more about the object that created the extreme tilt of Uranus. Using high-powered computers, the team were able to show the evolutionary history of the planet. In their simulation an object one to three times the size of Earth, postulated to be an icy body, impacted Uranus soon after its formation. Shigeru Ida remarked, 'Because the vaporization temperature of water ice is low and both Uranus and the impactor are assumed to be ice-dominated, we can conclude that the impact-generated disk has mostly vaporized.' This would eventually form into Uranus as, the researchers surmise, 'The disk cooled down enough for ice condensation and accretion of icy particles to begin.'[15] This new model was 'able to reproduce the observed mass–orbit configuration of the Uranian satellites. This scenario contrasts with the giant impact model for the Earth's Moon, in which about half of the compact, impact-generated, solid or liquid disk is immediately incorporated into the Moon on impact.'[16] The model is a tantalizing insight and goes some way towards explaining Uranus's unique tilt and rotation; it can be applied to exoplanetary systems that are far away from our own.

Future Missions

Remarkably, Voyager 2 is still the only mission to visit Uranus, back in 1986. There has never been a dedicated mission to Uranus or its neighbour Neptune, the only class of planet for which this is the case. The ice giants are the least explored category of planets in our solar system, and there is now an increasing desire to return to them. Fundamental questions remain unanswered about these two planets, including their origins and the processes that drove their formation. How can two planets with a common origin follow such different evolutionary pathways? Because of its differences due to its tilt, it has been suggested that Uranus has become an atypical ice giant, whereas Neptune is archetypal and hence a better paradigm for an ice giant orbiting another star.[17] The motivation to return to these planets is driven by the desire to answer ongoing questions about their ice-rich internal structure, unique magnetospheres, unusual atmospheric dynamics and the viability of life on their moons. As astronomers discover an ever-increasing number of giant ice worlds orbiting distant stars, this class of planet is no longer a rarity.

The need to return has been further heightened by the outstanding success of the ongoing NASA New Horizons mission. Launched in 2006, the probe conducted the first ever flypast of Pluto in 2015. The striking images received from New Horizons caused astonishment among hardened scientists and reproduced the wonder that the Voyager images had evoked thirty years earlier. Pluto, its moons and the Kuiper Belt object Arrokoth revealed surface details in colour for the first time. Regarding the Pluto flypast, Alan Stern, the New Horizons principal investigator from the Southwest Research Institute, Boulder, Colorado, said,

> It's clear to me that the solar system saved the best for last! We could not have explored a more fascinating or scientifically important planet at the edge of our solar system. The New Horizons team worked for 15 years to plan and execute this flyby and Pluto paid us back in spades![18]

Artist's impression of the Space Launch System on the launchpad, 2019.

Several follow-up missions to visit Uranus and Neptune have been proposed over the past few decades, but so far none have been approved for development. These include the NASA OCEANUS (Origins and Composition of the Exoplanet Analog Uranus System) orbiter concept. Additionally, the European Space Agency (ESA) proposed the Pathfinder, ODINUS and MUSE (Mission to Uranus for Science and Exploration) mission concepts. Pathfinder and ODINUS have since been abandoned, but MUSE is still under consideration for implementation.

NASA OCEANUS was proposed in 2016 as part of the New Frontiers programme. This spacecraft would enter Uranus's orbit in 2041 and then study the interior structure, magnetosphere and atmospherics in detail, in a way that a flypast could not. Launched in 2030 the mission would take eleven years to arrive at the system and cost an estimated U.S.$1.2 billion.[19] One challenge is that the orbital parameters of the solar system mean there are few opportunities for gravity assists beyond Jupiter during the 2020s and 2030s. Launch windows are short and the mission would require two gravity assists around Venus to reach the outer solar system, the first in November 2032 and the second in 2034. Once in orbit around Uranus, it would perform a minimum of fourteen orbits, all of them highly elliptical. A second possible option considered by NASA was suggested as part of the Planetary Science Decadal Survey: 2013–2022. The concept would launch an orbiter and an atmospheric entry probe with the newly commissioned Space Launch System (SLS) to propel the payload to the planet. The new core stage provides considerably more lifting power and

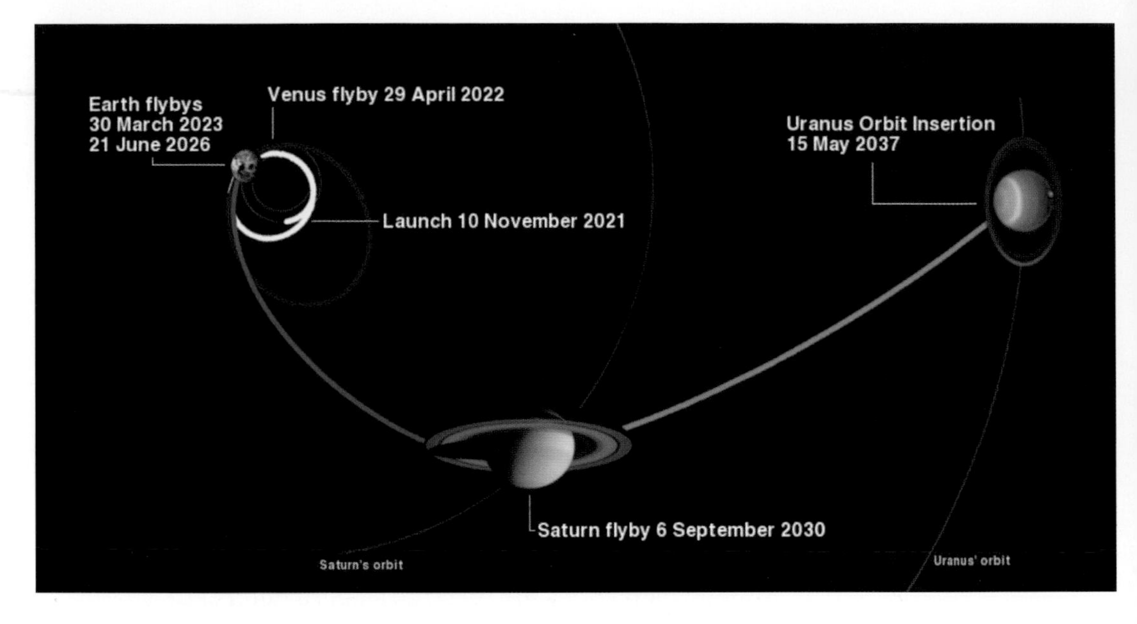

The proposed trajectory of Uranus Pathfinder.

lets long-range missions perform in a more cost-effective manner, opening up journeys to the outer solar system.

In 2010 a dedicated mission called Uranus Pathfinder was submitted to ESA's Cosmic Visions programme. The mission was highly rated and while it reached the final shortlist of eight possible missions, it was unfortunately not selected. If it had been chosen, the mission would have launched in 2022 and would have arrived at Uranus in 2037. One of the issues with sending probes to Uranus was explained by the principal investigator of Uranus Pathfinder, Chris Arridge of University College London: 'Both the European and American sides are convinced that an orbiter is needed rather than a flyby. But then costs rear their ugly head.' Even back in 2011 Arridge realized there was a growing need to return to this planet, saying, 'The project is by no means dead. Interest is building to go back to Uranus.'[20]

A second mission, ODINUS, was also proposed as part of ESA's Cosmic Visions programme. This mission would visit both ice giants, with an orbiter and a probe. Two orbiters, named Freyr and

Freyja after the twin gods of the Norse pantheon, would be launched in 2034. The idea behind this mission raised the bar in what was considered possible. It has now been shelved, but the design and idea behind it show how the boundaries of space travel can be continually pushed forward with new ideas and concepts.

MUSE is a European Space Agency mission to Uranus. Proposed for launch in 2026, it would take the spacecraft sixteen years to reach the planet, finally arriving in 2044. Consisting of an orbiter and probe, this mission would gather information on the planet's rings, moons, interior, atmosphere and magnetosphere. The orbiter would undertake 36 orbits around the planet, lasting two years before moving to an orbit around the moons for another year. There would be nine flypasts of the moons Miranda, Ariel, Umbriel, Titania and Oberon.

Although it is still in the proposal stage, a number of instruments could be carried on board, including spectrographs, magnetometers, electron sensors and high-resolution cameras. As with all space missions the instrumentation would be as comprehensive as could be, with weight restrictions and cost considerations taken into account. MUSE is still regarded as the best opportunity for a mission to Uranus. It will cost in excess of €1.8 billion.[21] On arrival it will need to conduct a wide range of studies as the desire to launch another mission may not arise for a long time afterwards. The increased interest in Uranus's moons and the possibility that the five largest moons may have subsurface oceans, thus making them possible candidates for life, is just one of the reasons this mission may be funded. The next chapter will look at these individual remarkable worlds and the other moons surrounding the planet.

FOUR

Moons of Uranus

'Sweet Moon, I thank thee for thy sunny beams;
I thank thee, Moon, for shining now so bright.'
BOTTOM AS PYRAMUS, *A MIDSUMMER NIGHT'S DREAM*

Centuries before Uranus's satellites were discovered, William Shakespeare wrote many times of our own Moon. It seems fitting that almost all of the moons around Uranus are named after characters in his plays, the final few taking their names from the literary works of Alexander Pope. John Herschel chose to name the moons after the king and queen of fairies, Oberon and Titania from Shakespeare's *A Midsummer Night's Dream*; the gnome, Umbriel, from Pope's *The Rape of the Lock*; and the sylph, Ariel, from the same poem (this name also appears as a character in Shakespeare's *The Tempest*). The naming is unusual, as moons within the solar system are usually called after beings from ancient mythologies.

Discovery

On 11 January 1787 William Herschel once again turned his telescope to Uranus. Nearly six years had passed since he had discovered the planet and it was considered to be a body orbiting alone in the outer depths of our solar system. During that evening's observation he discovered two bodies orbiting Uranus, the first two moons of the 27 now known.

John Herschel, 1846.

William's discoveries would become known as Titania and Oberon. After finding the first two moons, William would continue to observe Uranus and on 18 January and 9 February 1790 he reported seeing two more moons. They were followed by another two observed on 28 February and 26 March 1794, thus six in all. Something is amiss here, as we now know there are only five large moons orbiting Uranus. The original two were observed on a further thirty occasions before 1798, but the others were missing. Soon all the moons became too dim for Herschel to see through his telescope as the whole planet moved into an unfavourable observing position.

A Return to the Moons

William's son, John Herschel, returned to the observation of the moons of Uranus in 1828. John's observations enabled the determination of the moons' orbital size, inclination and eccentricity. An excellent observer in his own right, John was to report that he could not see four of the satellites that his father claimed to have observed.

No other astronomer was able to confirm these four new moons, but their presence was taken as fact and the spurious idea that Uranus had six moons appeared in a number of textbooks. British mapmaker James Wyld added them to his diagram of the solar system dated 1841, with Uranus still called Herschel. Denison Olmsted (1791–1859), an American astronomer who wrote the 1839 *Astronomy Textbook* for Yale University, had his doubts, and remarked:

> Uranus is attended by six satellites. So minute objects are they that they can be seen only by powerful telescopes. Indeed the existence of more than two is still considered as somewhat doubtful. These two offer remarkable, and indeed quite unexpected and unexampled peculiarities. Contrary to the unbroken analogy of the whole planetary system, the planes of their orbits are nearly perpendicular to the ecliptic, being inclined no less than 78° 58′ to that plane, and in these orbits their motions are retrograde; that is, instead of advancing from west to east around their primary . . . they move in the opposite direction.

To which a footnote was added: 'A third satellite of Uranus is said to have been recently seen in Munich.'[1]

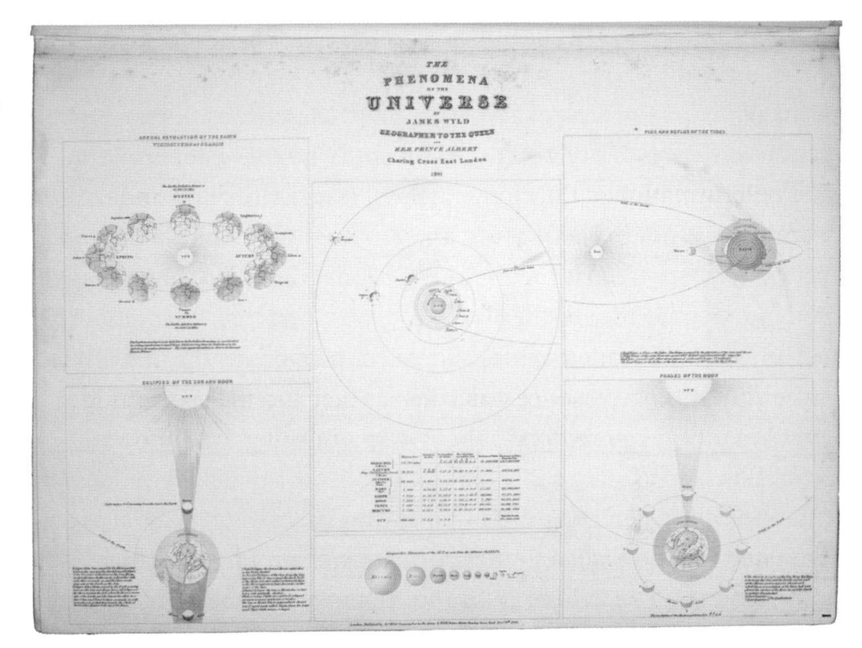

Drawing of the solar system in which Uranus (still called Herschel) has six small satellites. Taken from James Wyld, *Phenomena of the Universe* (1841).

Other Major Moons

In 1851 William Lassell reported his discovery of the next two major moons, Ariel and Umbriel. Writing about this to the Royal Astronomical Society, he explained:

> After a ten-day painful exercise of patience, I obtained last night, still under great disadvantages, another view of the two new satellites of Uranus. The sky was so generally cloudy and always hazy through the whole of last night, that I could get no absolute measures (save one) of either position or distance of either even of the bright satellites; though I managed to get a view of both the close ones, and also as good estimations of their situations with respect to the bright ones, as convinced me that no great error of position could exist.[2]

In this letter he was to confirm a number of earlier observations of Ariel and Umbriel, first made by himself and then independently by Otto Wilhelm von Struve (1819–1905) from as early as 1847. Lassell reported orbital periods of just 2.506 and 4.150 days, very close to the current values of 2.52 and 4.14 days. It was clear that these two moons were not either of the missing moons William Herschel had previously reported, as their orbital periods were much shorter. Adding these to the six already observed by Herschel, the number of moons around Uranus was considered to be eight by the mid-1850s, although four had yet to be confirmed. There was real confusion over the number of moons in the system, and diagrams of Uranus from this period often show the planet being orbited by seven moons. One such example is in the textbook *First Steps to Astronomy and Geography* published in 1828, where seven moons are clearly shown orbiting what was, at that point, the outermost planet.

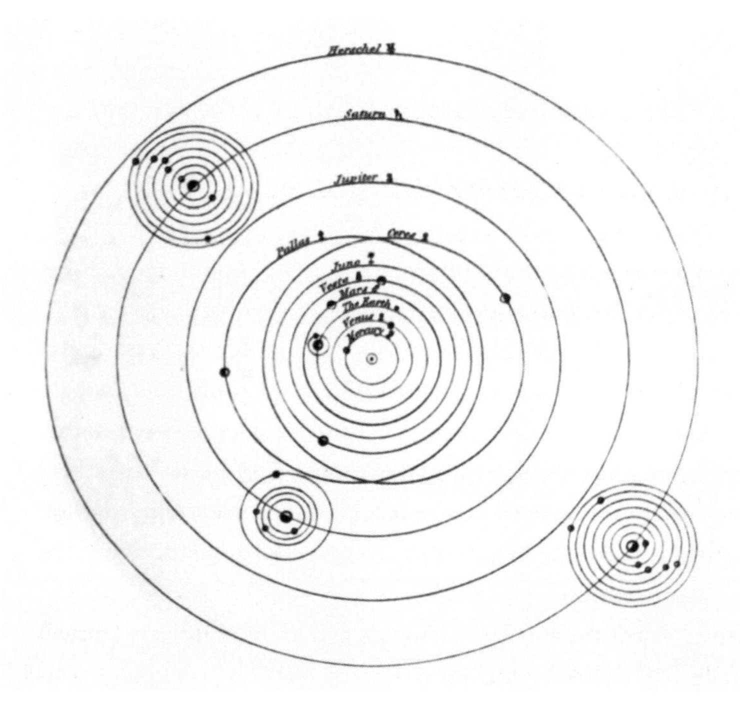

Seven moons orbit Uranus, from *First Steps to Astronomy and Geography* (1828), author unknown.

Number	Name	Discovery Date	Discoverer	Sidereal Period (Days)
1	Ariel	24 October 1851	William Lassell	2.52
2	Umbriel	24 October 1851	William Lassell	4.14
3		18 January 1790	William Herschel	5.89
4	Titania	11 January 1787	William Herschel	8.71
5		26 March 1794	William Herschel	10.96
6	Oberon	11 January 1787	William Herschel	13.46
7		9 February 1790	William Herschel	38.08
8		28 February 1794	William Herschel	107.69

William Lassell would have the last word on this matter and in 1853 he wrote that he was 'fully persuaded that either Uranus has no other satellites than numbers 1, 2, 4 and 6 of the list above [taken from G. F. Chambers's *Handbook of Astronomy* (1861)], or if it has they remain to be discovered'.[3] He was completely correct: only the named moons on the list existed. And what of Herschel's other moons? Well these are now thought to have been down to a combination of observer error and misidentified stars.

The other moons in the solar system up until the discovery of Uranus all orbited to within 35° of the ecliptic. By contrast, the moons of Uranus were orbiting perpendicular to the plane of the ecliptic. This led the famous Victorian astronomer Robert Stawell Ball to declare that 'The moons of Uranus seem to have got a twist, from some accidental circumstance for which we are not able to account.'[4] The discovery of Ariel in 1851 would be the last of the moons for one hundred years.

During 1875, Simon Newcomb (1835–1909), a Canadian American astronomer who worked at the United States Naval Observatory in Washington, DC, made a comprehensive set of observations of the Uranus system. He was determined to accurately describe the orbital elements of the four major satellites, designating them 'the most difficult well-known objects in the heavens'.[5] The moons were observed using a 26-inch refractor mounted at the U.S. Naval Observatory.

With this splendid telescope, Newcomb made numerous measurements. He proved that the moons moved around the planet in circular orbits, with a low eccentricity and the inclination that was in the same plane as the tilt of the planet. A German astronomer, Friedrich Emil von Asten (1842–1878), had used observations made by Struve to show this for just the moon Titania in 1872.[6] Newcomb's elements of the satellites of Uranus would set a benchmark that would be used for years to come.

William H. Steavenson (1894–1975), a British amateur astronomer, used his 30-inch reflector to carefully assess the brightness of the moons of Uranus in 1946–8, after a long period during which the moons had been largely overlooked.[7] He found that Ariel and Titania were the brightest with apparent magnitudes of 13.7, Oberon 13.8 and Umbriel a dimmer 14.5. Perhaps, surprisingly, Titania and Oberon had differences in brightness from night to night, which he concluded was a result of the rotation of the moons, during which they appeared alternately dark and then lighter. At that time Earth was positioned at a near right angle to the plane of sight and he was able to deduce that the two moons would be almost pole on to Earth, much in the way Uranus is in its orbit around the Sun. This was shown to be the case by measurements made by the Mount Palomar Observatory in 1950,

The 26-in. refractor telescope of the U.S. Naval Observatory which was used by Asaph Hall to find the moons of Mars.

which also found the apparent magnitudes of the largest moons to be (in order of brightness): Ariel 13.9, Titania 14.0, Oberon 14.1 and Umbriel 14.8.

Soon afterwards, on 16 February 1948, the Dutch American astronomer Gerard Kuiper (1905–1973) turned the McDonald Observatory's 82-inch telescope towards Uranus. He took a series of images to gauge the relative magnitudes of the four known moons. When he compared these images, it was obvious that another moon had been captured in the photographs. It was a 17th-magnitude object and orbited 123,000 km from the planet, being the innermost of the known moons at that time. In keeping with the literary nature of the names of the earlier moons, Kuiper suggested the name Miranda, taken from Shakespeare's play *The Tempest*. 'Oh brave new world', declared the character Miranda, and once Voyager 2 passed by this moon we can see how apt the phrase is.[8]

Voyager 2 and the Moons

Voyager 2's images of the five major moons were some of the highlights of the flypast. On 20 January 1986, the five major satellites were all imaged at ranges between 5 and 6.1 million km. Over the coming days the moons would be imaged in much closer detail, with craters and other surface features becoming clear. During the flypast Voyager was able to obtain improved values for the orbital characteristics of the satellites, particularly the orbital size, inclination and eccentricity of the moons.

Oberon

Voyager came within 660,000 km of the surface of Oberon. This allowed photographs to resolve features down to 12 km in size. Oberon is the outermost major satellite of Uranus, only being surpassed in size and mass by Titania. Oberon's orbit around the planet is almost circular and takes just 13.46 days to complete, the same time it takes to make a leisurely rotation on its axis. Therefore, one side of the satellite always faces the planet and it is thus tidally locked.

Along with the other major moons – Miranda, Ariel, Umbriel and Titania – it orbits within the planet's equatorial plane and as Uranus orbits the Sun at an almost perpendicular angle, effectively on its side, this means that Oberon is subject to extreme seasonal variations. It has long summers, with the poles being subjected to 42 years in the Sun and then complete darkness for another 42 years. When Voyager 2 photographed Oberon it was the southern pole that was in sunlight during its extended summer. The surface around the north pole was shrouded in darkness and is still a

The five large moons of Uranus in order of distance from the plane (from left): Miranda, Ariel, Umbriel, Titania and Oberon.

complete mystery to us. As only the southern half, about 40 per cent, of the surface was imaged by Voyager, caution is needed when drawing conclusions about its surface and composition, as the dark side of the moon could reveal different features which would lead to a revision of the geological history of the body. The illuminated side of Oberon shows a heavily cratered world. It has the most craters of all of Uranus's moons and this suggests that it is not geologically active, being a dead world almost since its formation.

There are so many craters that any new impact craters would almost certainly destroy existing ones. There has been a recycling of surface material, which has happened over extended periods of time. The dark-reddish surface has a number of bright ejecta rays radiating from some of the larger craters. Its brighter colour is relatively fresh water ice being thrown up onto the surface, due to the force of meteor impacts.[9] Craters have diameters of up to 210 km and the dark interiors of these craters is a topic for debate. It could come from a different substance below the surface, which would suggest the moon is in part differentiated with an icy crust on top of a rocky interior. A less likely alternative suggests a volcanic past where material has been pushed up from the depths of the interior, leaving darkened lava on the floor of the crater.[10] However, there is no evidence of recent volcanic activity or any internal heating. Another feature identified on Oberon's surface by Voyager 2 is called a 'chasma'. These are canyons similar to a rift valley, as found here on Earth.[11] They have been created during the period of formation, from the expansion and breaking of the crust due to internal cooling.

When Voyager 2 first measured the magnetosphere around the planet, it became evident that Oberon had one important dissimilarity from the other four major moons. Part of its orbit takes it outside the bounds of the magnetosphere, exposing it to the full onslaught of the solar wind. The result of this can be seen on the surface of the moon where there is a lack of dark material in the trailing hemisphere.

Oberon taken by Voyager 2 at a distance of 660,000 km on 24 January 1986.

Titania

Titania is the largest moon of Uranus, with a diameter of 1,576 km, which is only 50 km larger than Oberon's diameter. The two moons have much in common, both being formed from the accretion disc shortly after the planet itself. They are both composed of an equal amount of rock and ice, and the surface of both moons is predominantly old and pockmarked with craters. The largest crater on Titania is called Gertrude, named after the prince's mother in Shakespeare's *Hamlet*, and has a diameter of 326 km.[12] There are younger features such as a number of smooth plains which could show resurfacing due to lava flows from a process of cryovolcanism (occurring from volcanos found on icy bodies that erupt water, ammonia or methane). Titania has the distinction of having a number of large fault valleys running over its surface. In places, these faults or scarps run in parallel; between these is

Titania taken by Voyager 2 on 24 January 1986.

a narrow valley called a graben. A graben is created when the crust stretches and moves apart, forming distinctive depressions. The largest of these chasmata is called Messina Chasmata and extends for nearly 1,500 km, scouring a deep gorge through the surface of the moon.[13] It is thought to have been created by the expansion of subsurface ice.

Titania could have a tenuous atmosphere that is seasonally dependent. Any atmosphere would be composed of carbon dioxide, which would come from surface ices, as any nitrogen or methane would burn off in the sunlight and escape into space due to the weak gravity found at the moon's surface.

Umbriel

When Voyager imaged Umbriel, it found it was photographing another heavily cratered dead world, similar to Oberon – although, with a diameter of 1,169 km, it is a much smaller moon. During formation it may have been large enough to differentiate into a rocky core surrounded by a layer of ice made from water and carbon dioxide. It has the darkest surface of all of Uranus's satellites, reflecting only a fraction of the light it receives, and all the features on the surface have been named after dark spirits. Craters can be seen on the surface; they have bright central peaks but no impact rays, suggesting that the crust is made up entirely of dark material. One of the more distinct features on the surface is a ring of white material which is found in the impact crater Wunda (named after an Australian dark spirit). Its brightness makes it stand out against the black surface, but the reason for this is currently unknown. One possibility is that a large piece of carbon dioxide ice impacted the surface and deposited the bright material, or it could have come from a different subsurface material, exposed when the impact occurred. The canyons which criss-cross the surface are stresses caused during the formation, when the interior cooled and expanded, breaking the crust. These fault lines are currently unnamed, as the best image from Voyager 2 was unable to resolve them. The surface of Umbriel is predominantly old and little has changed since its formation.

Umbriel at Voyager 2's closest approach on 24 January 1986.

Ariel

Only 35 per cent of Ariel's surface was imaged during the Voyager 2 flypast. Remarkably similar to the largest major moon, Titania, Ariel's surface shows evidence of an active past. The photographs revealed the youngest surface features of all of the major moons. Geological activity on the surface has erased many of the older, larger craters. This activity is driven by tidal heating as a result of a more eccentric orbit in its past; the entire surface is thought to have renewed since its formation from the accretion disc, leaving it with the brightest surface of all of Uranus's major moons. The most striking features are the tectonic ridges or canyons that score across the surface of the moon. The longest of these is Kachina Chasmata, which is over 600 km in length and 60 km wide. This formed through the expansion of the interior from water ice as it cooled, late in the moon's evolution. This would have 'cracked' the surface,

The pitted surface of Ariel as imaged by Voyager 2 on 24 January 1986.

exposing a new canyon floor. Although this feature is impressive, much of the moon is covered by smaller canyons, a type of ridged terrain that would have formed as a result of similar stresses being placed on the surface during the cooling and expansion of the interior.

The youngest features are smooth plains. In one region they are observed on the floor of canyons.[14] Cryolava (lava made from ice) flow of up to 3 km in length is found, likely deposited in a similar way to a shield volcano.[15] Geologically active as recently as 100 million years ago, Ariel is a fascinating moon.

On 26 July 2006 the Hubble Space Telescope imaged Ariel in transit, moving in front of the planet. The image didn't resolve any features on the moon, but its shadow can be seen on the surface close by.

Ariel in transit across Uranus, taken on 26 July 2006 by the Hubble Space Telescope. Photographed in the near-infrared, the planet's oblateness and atmospheric banding are very apparent.

Miranda

Miranda is the least spherical of the major moons. This is due to its mass being at the limit of where its own gravitational forces will form a sphere. Miranda is also the least dense, made from 60 per cent water ice, with a smaller amount of rock. It orbits close to the equatorial plane of the planet, with an inclination of 4.34° – the largest inclination of all the major moons; the reason for this is unknown. It is possible this moon was subject to a large impact, smashing it apart, and that it went through the process of reassembly, although internal heating may have also contributed to the higher inclination and the small, less spherical shape. Due to the exceptionally close flypast distance of 29,000 km, Voyager 2 took excellent images of Miranda. Once again it was the southern hemisphere that faced the Sun, bathing the southern pole in sunlight. The imaged surface can only be described as complex. It remains unknown how a small moon could have the features seen on the surface, as it does not have enough internal energy to form them. Tidal heating could play a role. It is in a 3:1 resonance with Umbriel, and the flexing of Miranda could cause a 20 K increase in temperature.[16] Further resonances with other moons may add to this increase in temperature and subsequent development of the observed features. The surface is certainly diverse, with canyons, cliffs, scarps, craters and groove regions called 'coronae'. These features suggest intense geological activity, as some of them have overlaid previous surface features. One of the most unusual features is a unique patchwork pattern. By contrast, part of the surface is densely pockmarked with craters, and contains some of the oldest surface material in the Uranus system, suggesting a common birth with the other major moons. In common with Ariel and Titania, there is evidence of cracking in the surface due to the expansion of the interior when it was undergoing a period of cooling, and the resulting canyons extend up to 100 km in length.

Miranda imaged by Voyager 2 on 24 January 1986.

The craters on Miranda's surface, alongside Verona Rupes (bottom right of image). This fault scarp has a height of at least 20 km.

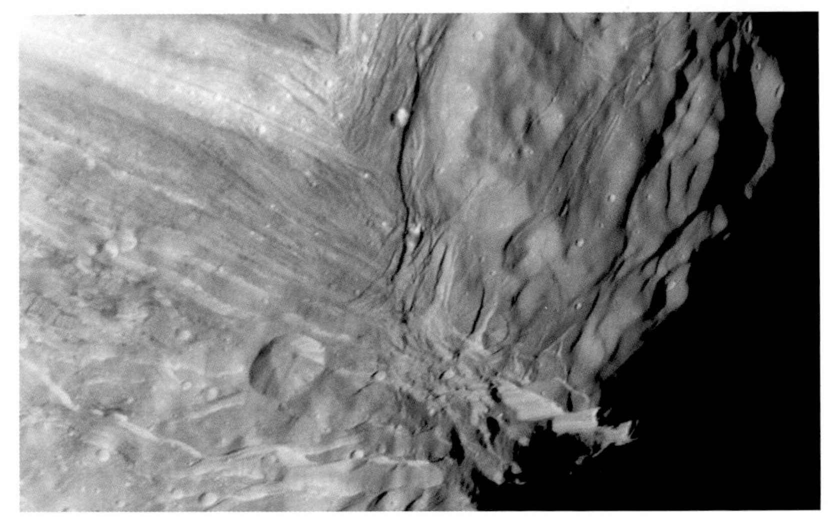

Location of the three coronae, Elsinore, Inverness and Arden, on the surface of Miranda.

Miranda has a uniform albedo and much of the surface has a rolling topography, but three regions found on the surface are visually unique.[17] These are coronae and have been named Elsinore, Inverness and Arden. Visually they are belts of bright and dark material and at their extent are banded by parallel ridges. This surface has little cratering and is a young terrain not seen elsewhere in the solar system, being less than 100,000 years old.

The coronae contain a number of fault scarps, notably Verona Rupes, which is the highest cliff in the solar system at an estimated 10–20 km in height.[18] As a result of the low gravity on Miranda, you might survive a fall from the cliff, as you would float gently downwards to reach the base. Coronae are unique to Miranda; they occur when fracture-controlled extrusions of warm ice escape from the interior. Elsinore also shows areas of young lava flow. Miranda's surface is a frigid 87 K and the material in the flow would have been too viscous to be a liquid, but nor would it be a solid. It is most likely to have been a mixture of ammonia and water which freezes at about 97 K and has a consistency similar to a thick, sticky substance like golden syrup. With an apparent magnitude of just 16.6, Miranda is too dim for almost all amateur telescopes, and what we know about its surface comes entirely from the Voyager 2 flypast. This moon, with its unusual geology, is one of the reasons that a return mission to Uranus is long overdue.

Inner Moons

There was excitement within the Voyager team when ten inner moons were discovered. They were photographed in the short time-window when Voyager was imaging the planet before it continued

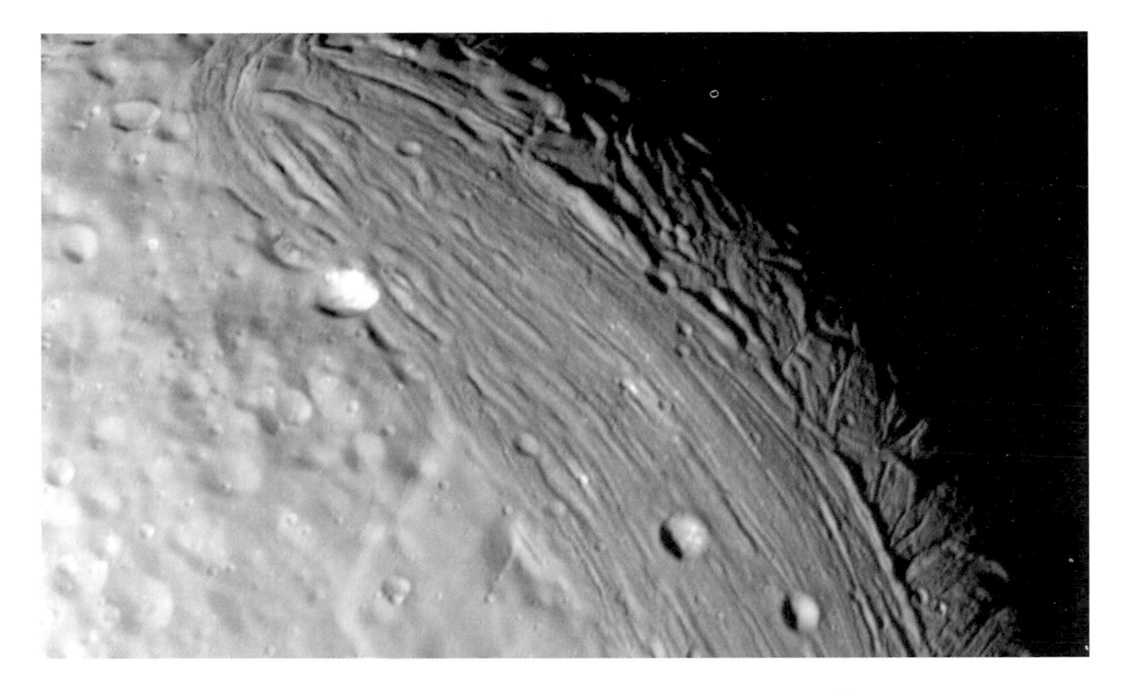

At least three types of geological styles on Miranda's surface.

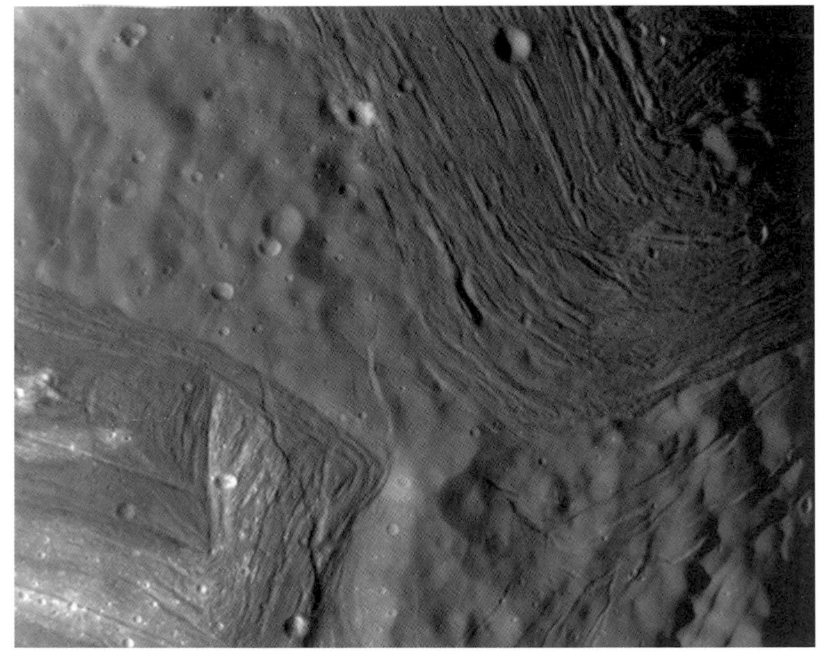

The fault scarps of the corona Elsinore (top right) and the white scoring of the Inverness corona (bottom left).

Uranus with three of the five largest satellites visible: Miranda, a speck (far right), Ariel (top right) and Umbriel (lower left). Imaged by Voyager 2 on 24 January 1986 at 74 million km.

on its journey to Neptune. In 1999 Erich Karkoschka examined the Voyager 2 images and spotted another moon; it would be called Perdita. The official recognition would have to wait until 2003, when the Hubble Space Telescope could image the moon to confirm its existence. Every year the Hubble Space Telescope images the Uranus system and it was not long before it discovered a further two inner moons, taking the total to thirteen.

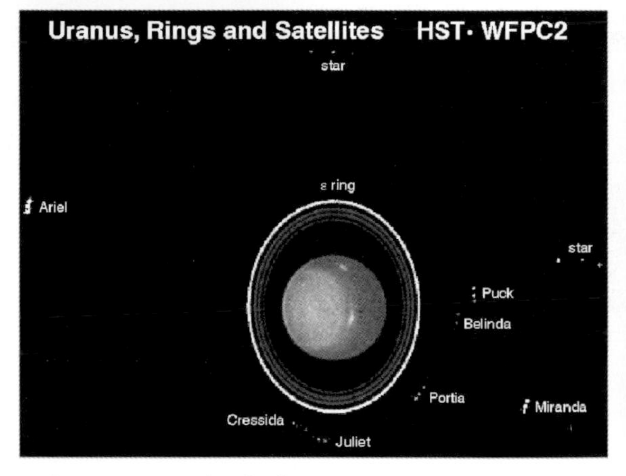

Five of the inner moons can be seen orbiting between the Epsilon ring and Miranda. Taken by the Hubble Space Telescope in 1994.

All the inner moons have small diameters and none are spherical. The most compact, Cupid, is only 18 km in diameter, whereas the largest, Puck, is 162 km in diameter. Puck's orbit is inside that of Miranda's, within the region of the μ ring. Almost spherical in shape, it is heavily cratered and although the resolution of the Voyager 2 image was poor, it was possible to make out a number of large depressions on the surface. Cordelia and Ophelia, the innermost of the known moons, are shepherd moons of Uranus's thin Epsilon ring.

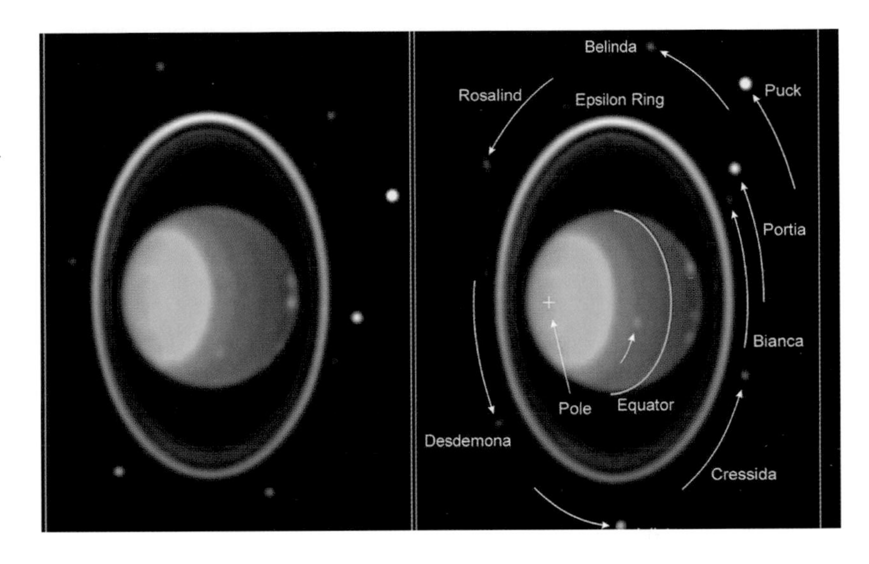

A family photograph of eight of the inner moons, captured by the Hubble Space Telescope's Near Infrared Camera and Multi-Object Spectrometer (NICMOS) in 1997. This is a very overcrowded region of space.

Outer Moons

Since 1997 nine distant non-spherical-shaped satellites, which are known as irregular moons, have been found orbiting Uranus. A team led by Brett J. Gladman using the 200-inch Hale reflecting telescope at the Palomar Observatory observed two small moons far outside the orbit of Oberon on 6 September 1997. The first was given the designation of S/1997 U1; later it was called Caliban after the half-human character in Shakespeare's play *The Tempest*. The second was initially called S/1997 U2 and later named Sycorax, after the mother of Caliban and a witch. Sycorax orbits twice the distance from Oberon as Caliban; they are both retrograde. Sycorax is currently the largest of all the outer moons: it has a diameter of up to 157+/-23 km.

Stephano and Setebos (originally designated S/1999 U1 and S/1999 U2) were discovered in July 1999 by another team led by Gladman, using the Canada-France-Hawaii Telescope (CFHT) on the 4,200-metre peak of Mauna Kea, Hawaii. The team took a number of images using a highly efficient charge-coupled device (CCD) camera, which collects large amounts of data; at the time, it was the world's largest electronic camera. These moons were twenty times fainter than the two found in 1997. Matthew J. Holman of the Harvard-Smithsonian Center for Astrophysics re-evaluated the images in the final week of July and discovered a third moon candidate designated S/1999 U3; it was later called Prospero, the protagonist of Shakespeare's *The Tempest*. The moon was only imaged on one night, 17 July 1999. This made it impossible at that time to know which direction the moon was orbiting around the planet. Later observations were to show that it was orbiting in a similar way to Sycorax, suggesting they had a common origin. 'The discovery of these irregular satellites is very important because it means that Uranus is not some oddball, but rather is just like Neptune, Saturn, and Jupiter,' declared Holman, as Uranus up until 1997 had no observed distant, and most-likely captured, moons.[19]

The smallest known moon to date was discovered by a group of astronomers led by Holman on 13 August 2001. Just 18 km in size, it was given the temporary designation S/2001 U1; later it was named Trinculo, after the jester in *The Tempest*. Ferdinand was observed by Holman's team on the same night and then reobserved on 21 September 2001. Initially the object was lost until Scott S. Sheppard imaged Uranus with the Subaru Telescope and found two small moons: the first was confirmed to be the lost moon, Ferdinand. The second would be a new moon called Margaret. Returning to search through the 2001 images in 2003, Holman and Gladman were to discover the moon Francisco. This tiny moon, with a diameter of 22 km, orbits the closest to Oberon out of all the outer moons.

Little is known about these worlds except what can be deduced from their orbits. All but one are in retrograde. Margaret being the odd one out; we will return to this enigmatic moon below. The inner four – Francisco, Caliban, Stephano and Trinculo – have

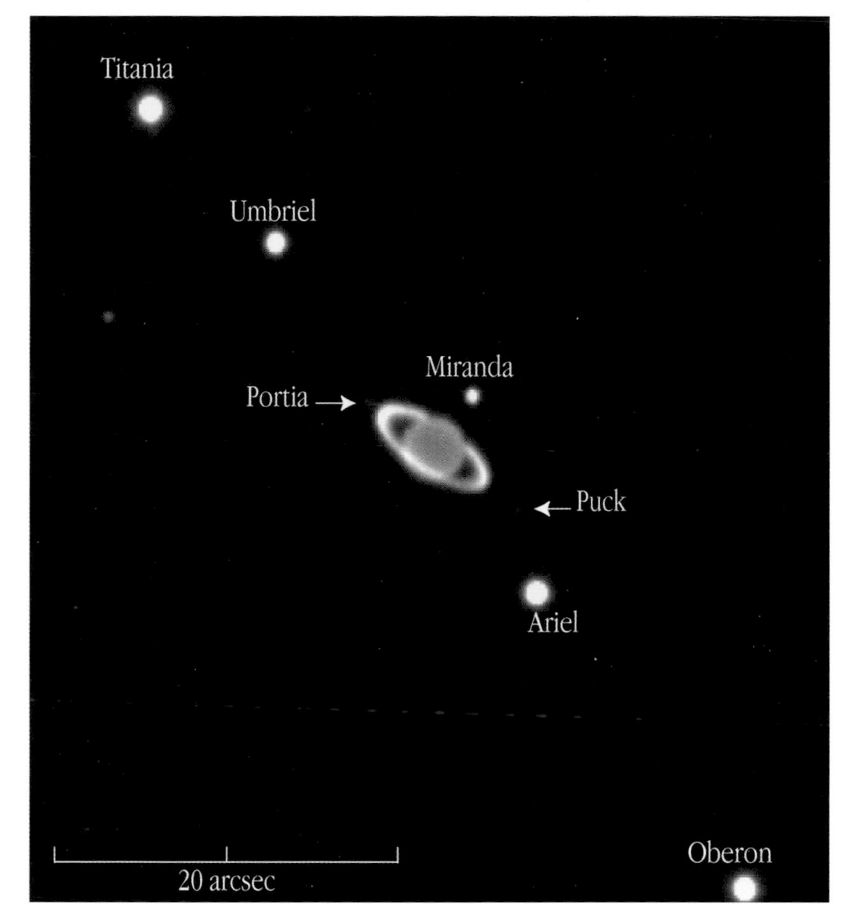

An image of the five major moons, along with Puck and Portia, taken in 2002 by the 8.2-m VLT (Very Large Telescope) ANTU telescope at the ESO (European Southern Observatory) Paranal Observatory (Chile).

moderately eccentric orbits, which are more stable than the outer four – Sycorax, Prospero, Setebos and Ferdinand – whose orbits are highly eccentric.

Margaret is a surprising oddball compared with the other outer moons. It is the only one with a prograde orbit, with a high eccentricity of 0.8, and it has a large 57° inclination. Its exceptionally high eccentricity puts the moon close to the limits of a stable orbit, which may eventually lead to its demise. It could undergo a phenomenon called the 'Kozai mechanism', where another moon would perturb it by its gravitational pull. In time this will set the moon on a collision course with one of the other inner bodies. The resulting crash will lead to the destruction of the moon or its ejection from the system altogether. This system is still dynamic and the lifetime of one of these moons is anticipated to be less than 1 billion years and perhaps as short as 10 million years.

Future Moons

The next generation of moons (if they exist) will be discovered using very large telescopes such as the Giant Magellan Telescope, which is being built at the Las Campanas Observatory in Chile. This telescope will have a resolving power about ten times more powerful than that of the Hubble Space Telescope. Currently it has been exceedingly difficult to resolve the small moons with a telescope. With the announcements of new moons being discovered around Jupiter and Saturn, Scott Sheppard explains why there have been none forthcoming for the outer gas giants: 'Those are so far away, we only know the limits for Uranus to 20 miles in size, or Neptune to about 30 miles in size.' This will surely not be the last word about the moons of Uranus.

FIVE

Neptune, Its Discovery and Pre-Voyager Observations

Neptune is the eighth planet from the Sun and the most distant in our solar system, at an average distance of 4.5 billion km (Pluto being defined as a dwarf planet since 2006). It orbits a cold region, receiving one-thousandth of the Sun's radiation compared with Earth. The nearest planet to Neptune is Uranus and at its closest this is over 1.6 billion km away. However, Neptune is not orbiting a part of space that is empty: its almost circular orbit is bounded by the Kuiper Belt, a region of space that contains comets and dwarf planets.

Origins

Neptune was formed alongside the other planets, moons, asteroids and comets from a protoplanetary disc made up of leftover material which encircled a newly born star, the Sun, 4.6 billion years ago. The early years of our solar system and planetary formation are still little understood. It is thought that the gas giants originally evolved in an orbit that was closer to the Sun but beyond the frost line, a region outside the current orbit of Mars. This is where volatile icy compounds (water, ammonia, methane, carbon dioxide and carbon monoxide) remain in a solid state as a result of the lower temperatures. After they were formed, it is possible that Uranus and Neptune swapped positions before they settled into their current orbits. In the early years of the solar system, there was more leftover material from the solar nebula and there would have been additional young bodies, many of which were destroyed either through collisions or amalgamation. The dynamic evolution of the solar system can be seen at Neptune where the moon Triton is believed to be a captured Kuiper Belt object. The addition of this object to Neptune's system would have been disruptive and it cleared many of its original satellites, thus creating the current system. Even now a number of the inner moons have been set on a spiralling path inwards and will eventually collide with Neptune.

Despite its large size, as Neptune is so far away from Earth (its average distance is 4.3 billion km), it is the only planet in the solar system that cannot be observed with the naked eye. At its brightest it will only reach a magnitude of 8 and hence had to be found using telescopes. The controversial story of its discovery will be familiar to many, as it is one of the most retold disputes in astronomical history. However, as William Sheehan states, 'the story, though remaining one of the most celebrated and oft-retold in the history of astronomy, is also one of the most complicated,' and what follows is a distilled version of its complex discovery.[1]

Discovery

In 1846 Urbain-Jean-Joseph Le Verrier (1811–1877) wrote to astronomers asking them to use their telescopes to look for an undiscovered planet. Observations of Uranus showed it did not follow a predictable path and, since the 1830s, scientists had wondered if another body outside its orbit could be the reason for this irregularity.

Le Verrier was an excellent mathematician who would become director of the Paris Observatory. In 1842 he took on the challenge of

Urbain Le Verrier, *c.* 1845.

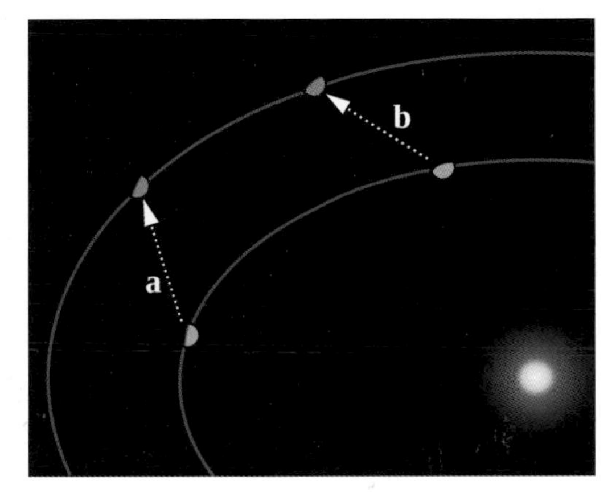

The perturbation of a planet on another planet is demonstrated in this diagram.

working out if another planet's gravitational influence could predict the deviation of Uranus from its orbit. His prediction not only gave the position of a new planet, but it offered the astronomers an idea of its apparent angular size, distance and mass. However, he needed help to search in the skies, as the Paris Observatory scopes were not up to the task, and he wrote to a number of eminent astronomers who had access to the world's best telescopes.

Unbeknown to Le Verrier, John Couch Adams (1819–1892), a young Cambridge graduate, had also been working on the same problem and had reached a similar conclusion about the planet and where to look for it. Adams had known about the Uranus problem for several years and during his time as a student at the University of Cambridge he had written himself a memorandum in which he described his intentions to try and undertake the exceedingly difficult calculations. With an interest in astronomy since his childhood in Cornwall, he is considered to have had one

Photograph of a young John Couch Adams. This portrait was taken at about the time of the discovery of Neptune.

of the greatest mathematical minds of the Victorian period. Adams wrote his memorandum in 1841, but it would take him some time to act upon his intentions. This was due to his ambitions to achieve the highest accolades in the Tripos exam and other commitments during his time as an undergraduate student. Having a drive to succeed at Cambridge from the moment he arrived, Adams would achieve the highest honour of becoming the Senior Wrangler (for attaining the highest overall mark in the Mathematical Tripos). At the end of his studies he returned to the Uranus problem and was able to calculate the position of the new planet. He sought to make the Astronomer Royal, George Biddell Airy (1801–1892), aware of his conclusions in 1845, leaving a note at Airy's personal residence after calling on him twice in a day without appointment. Little happened regarding Adams's calculations. A letter from Airy to Adams asking for more clarification surrounding the radius vector element went unanswered; the bustle of astronomical life took over once more and the prediction was put to one side by the Astronomer Royal.

On 1 June 1846 Le Verrier presented his memoir at a public meeting of the French Académie des sciences, with his conclusions about the perturbations of Uranus and the calculated position of the new body. This news filtered back to Airy, who recognized that he already had a similar solution in hand, from Adams. Both astronomers had used similar data on Uranus's motions and also a mean distance of the new planet based on Bode's law. It is therefore unsurprising that they reached similar results.

When Airy received a letter from Le Verrier asking for help to look for the new body, he was to mobilize and instruct James Challis (1803–1882) at the Cambridge University Observatory to undertake a similar search to what was occurring elsewhere, saying, 'I am asking . . . almost at a venture, in the hope of rescuing the matter from a state which is . . . almost desperate.'[2] There was no doubt about the importance of this search and during the summer of 1846 Challis was able to use the Northumberland 11.25-inch equatorial telescope to set about a methodical search. On 31 August 1846 Le Verrier produced a new memoir about the planet, detailing its mass and orbit and how he's had little success in getting the French to search for the planet. With no results from England, he wrote to Johann Gottfried Galle (1812–1910) of the Berlin Observatory asking for his help. In 1835 a new observatory building opened in central Berlin, close to what later became Checkpoint Charlie.

At the Berlin Observatory on the evening of 23 September 1846 Johann Gottfried Galle was observing through the 23-centimetre Fraunhofer refractor, assisted by a student, Heinrich Louis d'Arrest (1822–1875). Galle called out observed objects one by one while

George Biddell Airy, *c.* 1860.

Revd James Challis, date unknown.

d'Arrest checked their positions against the *Star Atlas Hora* XXI, one of a set of new star charts which had been made with the greatest accuracy in mind. Just after 10 p.m. Galle called out what appeared to be an 8th-magnitude star at RA (right ascension 22h 53″ 23.84″). To which d'Arrest answered, 'That star is not on the map.' Switching to a more powerful eyepiece to examine the object, Galle noticed that the 'star' in the eyepiece was not a pinpoint of light but had a disc. Remarkably, the two German astronomers had discovered Neptune in just 1° 03′ 06.7″ (the width of two full moons) from the predicted position. The doubtlessly elated astronomers immediately sought out the director of the observatory, Johann Franz Encke (1791–1865), who was holding his own birthday party that evening. On arrival

Heinrich Louis d'Arrest, before 1875.

Johann Gottfried Galle, 1880.

they persuaded him to leave the celebrations and return to the observatory to see the new planet for himself. The proclamation of a new planet could not be made that evening, as they needed to determine whether the object moved against the background of stars. It would take another night of observations for the Berlin observers to confirm its presence and they announced their discovery to the world on 25 September 1846. Galle and d'Arrest had discovered the planet on their very first evening. By contrast, the English astronomer Challis in Cambridge had adopted a conservative and laborious approach, scanning a larger region of the sky. When he heard news of the discovery, he was chagrined to find that he had observed and not recognized it as a planet on two occasions – 8 August and 12 August 1846.

Controversy

'We see it as Columbus saw America from the coast of Spain. Its movements have been felt, trembling along the far-reaching line of our analysis with a certainty hardly inferior to that of ocular demonstration,' proclaimed Sir John Herschel at the meeting of the 1846 British Association for the Advancement of Science.[3]

The failure to discover the new planet was to resonate deeply within the British astronomical community. At the time, Airy was criticized for his handling of the whole affair; much of this criticism has proved to be unfounded. The French were also frustrated by what they saw as a British attempt to steal the march on Le Verrier's achievements. Although the Royal Society awarded Le Verrier its highest accolade, the Copley Medal, this did little to improve the situation. Adams himself always acknowledged Le Verrier's priority and in 1846 said,

> I mention these dates merely to show that my results were arrived at independently, and previously to the publication of those of M. Le Verrier, and not with the intention of interfering with his just claims to the honours of the discovery; for there is no doubt that his researches were first published to the world, and led to the actual discovery of the planet by Dr. Galle, so that the facts stated above cannot detract, in the slightest degree, from the credit due to M. Le Verrier.[4]

As the priority issue was beginning to die down, criticism came from a new direction. Had Adams and Le Verrier been overly confident with their predictions? Was the fact that their results and the actual planet were just outside of 1° apart a 'happy accident'? This suggestion was made by Benjamin Pierce (1809–1880), who showed that Neptune was in close resonance with Uranus. His view was very unpopular among mathematicians, who pointed out that the planet had been predicted and found within almost a degree. Nonetheless, the newly discovered planet had marked differences from the orbital parameters that Le Verrier and Adams had suggested. It orbited the Sun at 30 AU, but Le Verrier's two solutions had predicted it would be far more distant. The first prediction from 1 June 1846 located the planet at 37.9 AU and the second from 31 August 1846 at 35 AU. Although modern work into the calculations undertaken by Hiu man Lai and Kenneth Young show that Adams and Le Verrier's calculations have merit, they found that Adams and Le Verrier had been able to determine the position of Neptune, as their use of the inverse perturbation method coincided with a time of special circumstances, when the two planets had shared the same heliocentric longitude (1822) and the perturbations were close to resonance.[5]

Naming the Planet

The new body clearly needed a name, and a number of candidates were put forward. 'Le Verrier' was one of the first to be suggested after the great predictor himself. The British opted for 'Oceanus', while Galle proposed 'Janus'. Le Verrier suggested Neptune early in the discussions, although he was not happy when the Bureau des longitudes adopted that name, and the symbol of a trident, for the planet.[6] He wrote in anger to Airy on 26 February 1847, having now accepted the idea of having the honour of a planet being named after himself:

> When the planet was discovered, it was proposed by the Bureau of longitudes to call it Neptune. I was not part of the Bureau at the time, and I did not charge it with this decision . . . I declared . . . to M. Arago that the Bureau was a little too hasty, and that I would specifically entrust him with the task of presenting to the Academy of Sciences whatever he judged to be most suitable. Since then I have had no further involvement in the matter.[7]

The matter was soon settled, as Airy wrote back saying the English would be adopting the name Neptune.

Pre-Discovery Observations

After Neptune's identification as a new planet, there was the inevitable question: had the planet been seen before? A number of pre-discovery observations were considered. Louis François Wartmann's (1793–1864) 'planet', recorded in 1831, did not fit Neptune's orbit and thanks to careless errors he made during his observations, it was most likely Uranus that he had observed. A second false trail was laid by Niccoló Cacciatore (1770–1841). He recorded in 1835 an 8th-magnitude object in Virgo, which, after a

A diagram of the solar system showing the name of Le Verrier instead of Neptune. It also labels Uranus as Herschel, 1849.

single night of observations, was unable to be recovered and was never viewed after its initial sighting. A possible early and oldest-known observation of the planet could have been made by Galileo Galilei (1564–1642) on 28 December 1612, just a few years after the telescope was first turned to the night sky. Galileo carefully marked it in his notebook as a star in close conjunction with Jupiter. Neptune's motion between the observations is thought to have been too tiny for Galileo to determine with his small telescope.[8]

The U.S. astronomer and mathematician Sears Cook Walker (1805–1853), in 1847, looked through 50,000 star positions in the star catalogue *Histoire céleste française* by Jérôme Lalande (1732–1807). He was able to find Neptune marked on the star map on two occasions. The planet had been recorded first on 8 May 1795 and then just two days later on 10 May 1795. The slight discrepancy in position between the two observations showed that the planet had moved along its orbit between the two days. When Walker re-examined this area of sky, he found that Lalande's star had now vanished, so he had in fact recorded the planet. Some of the earliest calculations had come within three weeks of its discovery. Walker's discovery of Lalande's two observations extended the orbital position of the planet to over fifty years before its discovery, allowing for the first accurate elements of the planet to be calculated. Adams had already extended the orbital baseline with the observations made by Challis and was able to release the first predicted orbit of the newly discovered planet. He gave a value of 30 AU. Soon afterwards other pre-discovery observations were identified, including those made by John Herschel.

The Orbit and Mass of Neptune

Walker's elements can be seen in the table below in comparison to the original predictions made by Le Verrier and Adams. Walker's values were far closer to current-day ones and he calculated the distance to the planet, its eccentricity, period and mass for the first time. Neptune's journey through the solar system takes it on an almost circular orbit; it has little eccentricity.

Comparison of Orbital Elements of Neptune from Le Verrier and Adams's Pre-Discovery and Walker's Post-Discovery				
	Modern Values	Walker	Le Verrier 31 August 1846	Adams Hyp II
Mean Distance (AU)	30.11	30.25	36.15	37.25
Eccentricity	0.00867	0.00884	0.10761	0.12062
Period (Years)	164.8	166.381	217.387	227.3
Mass (Earth = 1)	16.648	19.978	57.603	33.297

Neptune's mass was another surprise. Le Verrier and Adams had predicted a planet that would have been more distant and larger in size. The planet was smaller than expected and there was no easy way to determine its mass. Neptune is the third most massive planet in the solar system, being exceeded in its mass only by Jupiter and Saturn. It has a mass of 1.02413×10^{26} kg, which is just over seventeen times more massive than Earth, but only one-nineteenth the mass of Jupiter, the largest of the gas giants. It also has the fourth largest diameter in the solar system, after Uranus, Saturn and Jupiter. This makes it the smallest ice giant in the solar system, with a mean radius of 24,622 km – just less than one-third of the size of Jupiter but nearly four times larger than Earth.

A Ring?

Perhaps one of the first long-standing mysteries surrounding Neptune is the claim made by William Lassell that he observed a ring system. *The Times* published his letter about this discovery on 14 October 1846:

> Sir, On the 3d instant, while reviewing this object [Neptune] with my large equatorial, during bright moonlight, and through a muddy and tremulous sky, I suspected the existence of a ring round the planet; and on surveying it again for some time on Saturday evening last, in the absence of the moon, and under better, though still not favourable atmospheric circumstances, my suspicion was so strongly confirmed of the reality of the ring, as well as the existence of the accompanying satellite.[9]

Lassell became strangely quiet about the idea of a ring system when reporting later observations of the planet made through the same telescope. Nevertheless other eminent Victorian astronomers, including Challis, were also convinced they had seen evidence of appendages or luminous phenomena around the distant planet, which could be conceived as rings. Unfortunately, these early accounts of a ring system did not survive the test of time, and it did not present itself again to ground-based observers until many years later. These early observations of the rings were an illusion, as they were far too delicate to be seen from Earth in the telescopes available at that time. Considering these observations, Richard Baum (1940–2012) concluded that these early rings arose from optical limitations. In short, they were an observational artefact.[10]

In the 1860s William Huggins (1824–1910) pioneered spectroscopic techniques, gathering data on the spectrum of stars, galaxies and nebulae. In 1871 he made a series of photographic observations of Neptune in the blue-violet end of the spectrum. Huggins identified a number of absorption bands that had a close resemblance to those found for Uranus. He recognized hydrogen but some of the absorption lines were difficult to assign and their identification would have to wait sixty years. Vesto Slipher found that the mountain location and elevation of the Lowell Observatory was a perfect place to investigate further, as the location, which placed the observatory high in the Earth's atmosphere, increased the capabilities of a spectrograph at the infrared end of the spectrum. By 1933, when Slipher had made a quarter of a century of observations, he realized that even at high altitudes images in the infrared gave little to no trace and he correctly surmised that these longer wavelengths were absorbed by Earth's atmosphere. Working with Arthur Adel in 1934, Slipher was to correctly identify methane as one of the major components of Neptune's atmosphere, since he found the relevant absorption bands at 816μm, 802μμ, 683μm, 673μm and 584μm.[11] These were precisely the absorption lines that Huggins had struggled to identify.

Markings on the Disc

Observations of the planet made in the late nineteenth and early twentieth century showed a disc that was featureless. Edward Emerson Barnard (1857–1923) was a very gifted observational astronomer whose name has been immortalized for his discovery of the high proper motion (its apparent change in position over time) of Barnard's Star. In 1895 he was taking measurements of the planet's diameter and the position of Triton, using the 36-inch equatorial at the Lick Observatory, and reported no markings, even though the 'disk was fairly well seen'.[12] This was the world's largest refracting telescope from 1888 until the Yerkes's 40-inch refractor saw first light in 1897.

From 1897 Barnard made numerous observations of Neptune at the Yerkes Observatory through its new 40-inch refractor. He was extremely impressed with the refracting telescope, which he felt had an exceptionally fine object glass lens made by Alvan Clark & Sons on a mounting by Warner & Swasey Co. At that time it was the largest refracting telescope in the world. A total of 51 observations

The Yerkes's 40-in. refractor, in Williams Bay, Wisconsin, 1897.

of Neptune were made from September 1897 to April 1898, although poor weather limited the time he was able to use the telescope during the winter months and by the summer, the best season for observing the planet was effectively over. During the first few months there had been issues of broad beams of 'ghostly light' shining through the lens. Examining the tube, the team found that spiders had woven a number of large webs close to the objective lens inside the tube and as soon as these small creatures and their webs were removed, the object glass worked perfectly. Once more Barnard reported no disc markings during his time at the Yerkes Observatory.[13]

A dubious claim of planetary surface markings came from the astronomer Thomas Jefferson Jackson See (1866–1962), who observed the planet through the 26-inch refractor at the U.S. Naval Observatory in Washington, DC, about the same time as Barnard. See was originally considered an excellent observer, but he was becoming a controversial figure who fell out with many of his colleagues. He took the post at the observatory in 1899 after having been dismissed from the Lowell Observatory. During this time, his work became increasingly erratic and his observations of a triple star system published in the *Astronomical Journal* in 1886 came under scrutiny, eventually being dismissed by the wider astronomical community. See suffered a nervous breakdown in 1902 and, having been marginalized by his peers, he was transferred from the U.S. Naval Observatory to work at the naval shipyard at Mare Island, California. His observations of markings on the planet are considered doubtful.[14] He spent the later years of his long life attempting to discredit Einstein's theory of relativity, with little success.

Early Ideas of Neptune's Structure

Because of their similar size and mass, Uranus and Neptune have often been considered to be twin planets, in much the same way as Venus and Earth. Even with the lack of information about these distant planets in the late Victorian period it was known that they had many divergent features as well as similarities – above all, the orbital orientation of Uranus and its known satellites, which was in marked contrast to Neptune. Robert Mann (1817–1886) proclaimed, 'The remote Neptune is almost certainly as much the twin of Uranus in its general features as Saturn is of Jupiter.'[15] Neptune's composition intrigued scientists as soon as the planet was discovered. Early ideas suggest that the planet was a failed star, which was completely gaseous, since all the gas giants were assumed to be hot. This was disproved by Sir Harold Jeffreys, in 1923, who theorized that there was a gaseous layer over a solid core and that Neptune, along with Uranus, was an icy cold body.[16]

By 1947 Robert Wildt (1905–1976) suggested the planet had a rocky core surrounded by a layer of water ice, with a gaseous atmosphere.[17] The core's diameter was 19,300 km, the ice layer 9,600 km and the atmosphere a modest 3,200 km. Just a few years later, in 1951, William H. Ramsey would describe the planet as being composed mainly of water, methane and ammonia, and thus markedly different to Jupiter and Saturn, which are composed mainly of hydrogen and helium.[18] This is a model that has stood the test of time and is still favoured today.

Up to the mid-twentieth century, surface markings on Neptune had not been reported by any trustworthy source. Then on 17 April 1953 the astronomer Thomas Cragg (1927–2011) observed a bright equatorial band and dusky poles. These observations were made from the Mount Wilson Observatory through the 60-inch reflector using 1,000× magnification. Cragg was a skilful observer who spent many years of his working life at the observatory in California, predominantly studying solar observations. This was the first hint that Neptune may have an active atmosphere.

First Images of Clouds

Due to the huge distances involved and the relatively small size of Neptune, early Earth-based observations of the planet were limited. Visually, in even the largest telescopes, it can be observed as little more than a tiny blue-green disc, upon which darkish bands might be suspected. Uranus's and Neptune's discs were considered to be so blank they were once used as albedo standards in the pursuit of establishing the existence of small variations in the solar constant.[19] Infrared astronomers would succeed in establishing the existence of weather changes on Neptune in the late 1970s.[20] Improvements in ground-based equipment and techniques allowed astronomers to start observing features within the upper atmosphere, and by 1979 the first clouds were imaged using charge-coupled devices (CCDs) attached to the 2.5-metre du Pont Telescope at the Las Campanas Observatory, Chile. It was suggested by American astronomer Bradford A. Smith that the clouds were most likely in the higher atmosphere and composed of methane ice or photochemically produced particulates. They rotated in the same direction as the planet and were not in retrograde. Unfortunately the optical limitations of the equipment and the temporal coverage of the clouds meant it was not possible to measure the rotation period of Neptune.

In 1982, as Voyager 2 passed Saturn on its journey to the ice giants, Smith and American astronomer Harold J. Reitsema explained, 'Observations of the minute disks of Uranus and Neptune from the surface of the earth present a major challenge to any observatory site and require the most advanced techniques in optical imaging instrumentation.' And they went on to conclude that the discs of Uranus and Neptune are probably featureless in visible light, that is, they do not show the same type of banded, atmospheric structure which is so familiar a characteristic of the images of Jupiter and Saturn.[21] Smith and Reitsema, along with American astronomer Stephen M. Larsen, continued to observe the planet using the 1.54-metre Kuiper Telescope reflector at the Catalina Station/Steward Observatory of the University of Arizona. In 1983 they were delighted to confirm that Neptune had cloud structures and remarked that 'the Neptune images are characterized by several patches of haze located at the mid-latitudes in both northern and southern hemispheres . . . Neptune appears to have cloud structure.'[22] From this distance the markings were vague and it would take the Voyager 2 flypast in 1989 to give an improved view of the surface of the planet.

Possible Rings?

The early observations of rings had been discounted and modern investigations into the rings were needed. Uranus's rings had been observed using the occultation method in 1977. Would Neptune fade and then brighten again during a similar occultation? One such investigation proved disappointingly inconclusive and astronomers suggested the rings may be incomplete. David A. Allen, of the Anglo-Australian Observatory, attempted to investigate the question in the infrared range of the spectrum using the 3.9-metre Anglo-Australian Telescope at Siding Springs Observatory; a similar venture had provided excellent views of the Uranus ring system.

The Las Campanas Observatory in the Atacama region, Chile, where the first images of clouds around Neptune were made.

By 1983 he conceded that he had failed to find a ring system around Neptune, writing, 'The results were disappointing. The planet is too small to see clearly and shows no hint of a ring.'[23] James L. Elliot and Richard A. Kerr, who were part of the team that had discovered Uranus's rings, said, in 1984, that 'if Neptune has rings, they will almost certainly not be discovered from the ground.'[24] They did not have too long to wait, as Voyager 2's flypast of Neptune was scheduled for the summer of 1989.

SIX

Voyager 2 Flypast of Neptune

In 1988 the English amateur astronomer and popular broadcaster Patrick Moore (1923–2012) commented on Neptune, saying he had 'done his best to index all the published papers of any importance, and the fact that I could do so in less than a hundred and fifty pages shows how meagre our knowledge was at that time. Since the Voyager mission, things have changed, and the published papers amount to many thousands.'[1]

Before its approach to Neptune, scientists did not want to build expectations up too high. Previous photographs taken from Earth had shown a featureless disc and after the flypast of Uranus and the resulting images of its unvarying surface, scientists anticipated a similar situation for Neptune. It had been theorized that due to the lack of sunlight reaching the planet, the disc would be placid and certainly no more interesting than the surface of Uranus. Voyager 2 arrived at Neptune in the summer of 1989; it was the first and, at the time of writing, the last flypast of the planet. The images beamed back to Earth were not only scientifically valuable but visually stunning.

Seven months before the spacecraft was scheduled to make its closest approach, in January 1989, the cameras turned to the planet and imaged it from a distance of 310 million km. These images gave the Voyager team the first indication that the surface was anything but placid; there were large variations in tone that hinted at what was to come. In the middle of the nineteenth century, the English clergyman and amateur astronomer Revd Thomas W. Webb (1807–1885) postulated, 'Who can say how grand a spectacle this inconspicuous globe might present on nearer approach?'[2] This question had lingered, unanswered. Now the scientists in the Voyager team were about to find out. Tom Spilker, a member of the Voyager 2 radio science team, recalls his excitement on the approach: 'I got this overwhelming feeling inside, as if I was standing in the bow of Captain Cook's expedition into the Gulf of Alaska for the very first time. We were going to places where no one had ever gone before – we were explorers.'[3]

Problems with the Spacecraft

The trip to Neptune had not been without its problems. On arrival the Voyager team would face a number of new challenges. Given the proposed longevity of the lifespan of the mission it is perhaps surprising that the peak performance of the spacecraft had only been designed to last five years and it was already twelve years into its journey. During this time the spacecraft had encountered problems that needed to be resolved before arrival. The azimuth of the scan platform which slewed the camera jammed as the spacecraft had flown past Saturn. The platform had been designed to enable the camera to track and slew at rates of about one degree a second. The British astronomer Garry Hunt, a member of the imaging team, explains:

> One night we came back from dinner and we got a phone call. And they said to us, 'All these pictures are dark space.' And we looked at each other and said, 'Oh we're calibrating. We're just taking calibration pictures.' We put the phone down and said, 'What the hell has happened?'

This was a nervous time and Hunt recalls they went through 'three days of hell' before the spacecraft was recovered. 'You can't call out the AA man to help you. You think, ah, we have the engineering model in von Kármán [building], and I checked all the ways it could work.'[4] Thankfully, after a very anxious wait, the scan platform started to move again but at a much-reduced rate of 0.08°/sec.

Another complication was the distance of the planet from the Sun. Neptune receives less than one-thousandth of the intensity of light that we do here on Earth; longer image exposures were needed to capture detail on the surface, its delicate ring system and moons. The images had to be at least 96 seconds in length and at times multiple images would need to be taken and then stacked together. There was a real concern that the motion of the craft itself could leave the team with blurred results. A number of modifications were made to the spacecraft, including the altitude control system, to compensate for this smearing. Then the team devised a method to take images in a specific way to stop blurring. They would either track the object by rotating the whole craft or follow the object by turning the scan platform, when possible, at its now fixed reduced speed. Transmitting the images back to Earth was also a challenge because of the distances involved. The strength of the radio signal decreased with every planet Voyager reached and by Neptune the transmissions needed to be compressed. Although 150 million km further out than Uranus, the data was transmitted at the same rate – 21.6 kbps. Instead of all the pixel data being transmitted, only the difference in the brightness levels of each pixel was sent, which reduced the data load by 70 per cent. At the same time, scientists were making improvements to the receivers here on Earth. More powerful receivers were built, including NASA's Deep Space Network, as well as the Very Large Telescope in New Mexico. These receivers permitted Voyager's weak signals to be converted into awe-inspiring images of the planet accompanied by an exceptional amount of important scientific data. The achievements of the engineering and science teams behind these improvements and work-arounds cannot be overstated.

Storms and the Climate

From the first image in January 1989 the scientists in the Voyager 2 team were enthralled that Neptune was not bland at all and had a number of cloud formations. Ralph McNutt, who worked in the plasma data team for Voyager 2 and later on New Horizons, exclaimed, 'We had been lucky enough to be in the right place at the right time with the right team, and this was the first and only opportunity we would have for a long time for an up-close and personal view with Neptune and the outer parts of our solar system.'[5]

By April the cameras on Voyager 2 were imaging Neptune at a distance of 208 million km. The spacecraft took three pictures ninety minutes apart. These images and those taken over the upcoming months would show the movement of the dynamic atmosphere. Excited scientists started to name the features they saw on the surface; the largest one was dark in colour and was called the 'Great Dark Spot' as it was reminiscent of Jupiter's 'Great Red Spot'. In the mid-1980s photometry of the planet had shown an increase in brightness in Neptune; between then and the year 2000, there was an increase of 11 per cent. This had been attributed at the time to a vague 'transient atmospheric discrete feature'. Neptune's visual magnitude

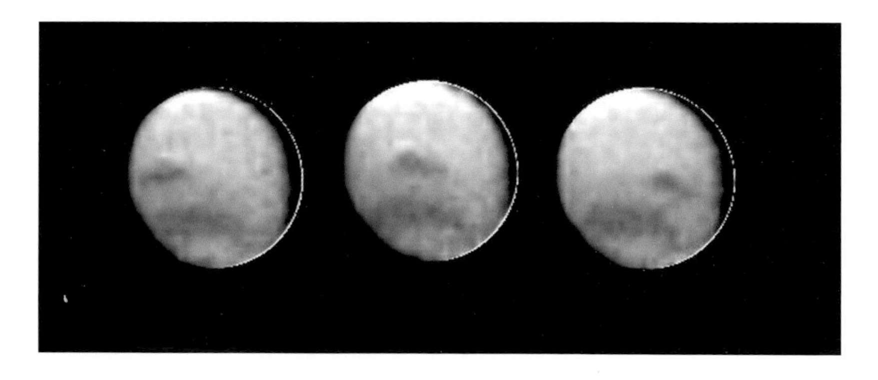

Neptune taken by Voyager 2 on 3 April 1989, from a range of 208 million km. Three images taken 90 minutes apart show features on the planet, including the 'Great Dark Spot'.

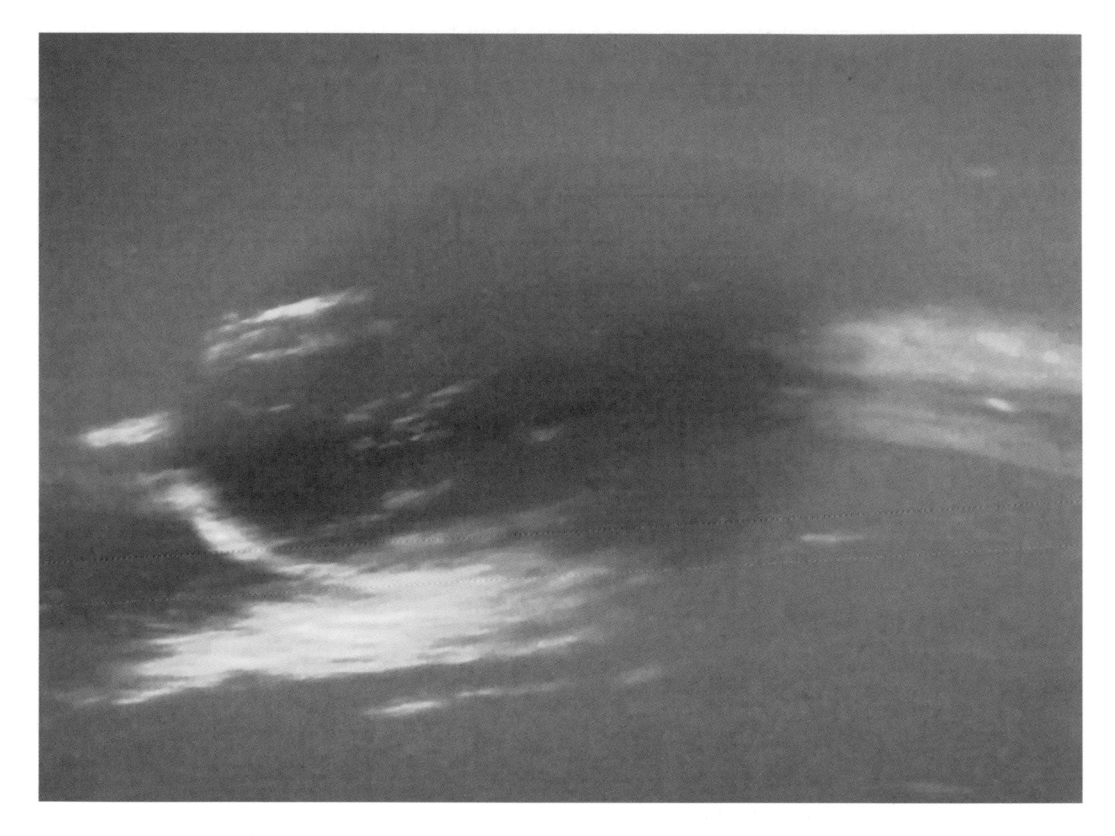

The Great Dark Spot of Neptune.

was getting brighter because of the increase in cloud bands, which was a result of seasonal changes and tilt of the planet.[6]

As the spacecraft approached the planet, the southern bright feature seen in remote images was identified as a patch of cirrus cloud of methane, skirting along the southern edge of a tremendous anticyclonic storm system, the Great Dark Spot. This was the largest feature on the whole disc, with a width equal to that of Earth. It inhabited almost the same latitude (20°S) as Jupiter's Great Red Spot, which was also an anticyclonic storm. At the edges of the storm the wind speeds were some of the greatest found anywhere in the solar system, reaching a velocity of 2,100 km/hr.

The dark spot resulted from a vortex in the upper methane layer. Within this vortex, white clouds were formed through updraughts in

Vertical relief in the atmosphere of a giant planet. Cirrus clouds project shadows onto the blue methane cloud deck below.

The 'bull's-eye' Dark Spot' image, 24 August 1989.

the atmosphere. These are thought to be similar to cirrus clouds on Earth, except that Neptune's cirrus clouds are made from ice crystals of frozen methane. These features can last for days on Neptune, rather than the hours that they exist within our own atmosphere. Voyager 2 imaged similar clouds in the northern hemisphere near the day-night terminator, where thin silvery wisps of cloud had cast measurable shadows on the sea-blue ocean below.

At first it was thought the Great Dark Spot would live as long as the corresponding Red Spot on Jupiter. It had been present before the Voyager mission, having been registered as a 'transient atmospheric discrete feature' in the observations made using photoelectric photometry in the late 1980s. However, by late 1994, when a team led by Heidi B. Hammel imaged Neptune with the Hubble Space Telescope, it had disappeared. This initial dark spot was just the first; there have now been several more imaged by the Hubble Space Telescope and these will be discussed in more detail in the next chapter.

Smaller white clouds were imaged rotating around the planet; the formation was nicknamed the 'scooter'. Its rotation speed was faster than the Great Dark Spot; it 'scooted' round the planet. Beneath this feature was another surprise, a second dark spot. Imaginatively named 'Dark Spot 2' it was smaller than the Great Dark Spot but it was equally dynamic. As Voyager approached the planet, the dark core became bright. It is believed to rotate clockwise, which is opposite to the vortex of the Great Red Spot on Jupiter, which rotates anticlockwise.

On 25 August 1989 Voyager 2 made its closest approach to Neptune. It came within 4,905 km of the planet's upper cloud deck at a point close to the north pole. Using the planet's gravitational pull the tiny spacecraft sailed towards Neptune's intriguingly large satellite, Triton, and passed within 40,000 km just six hours later.[7] It was three and a half years since Voyager 2 had passed Uranus and just over twelve years since it had set off from Earth.

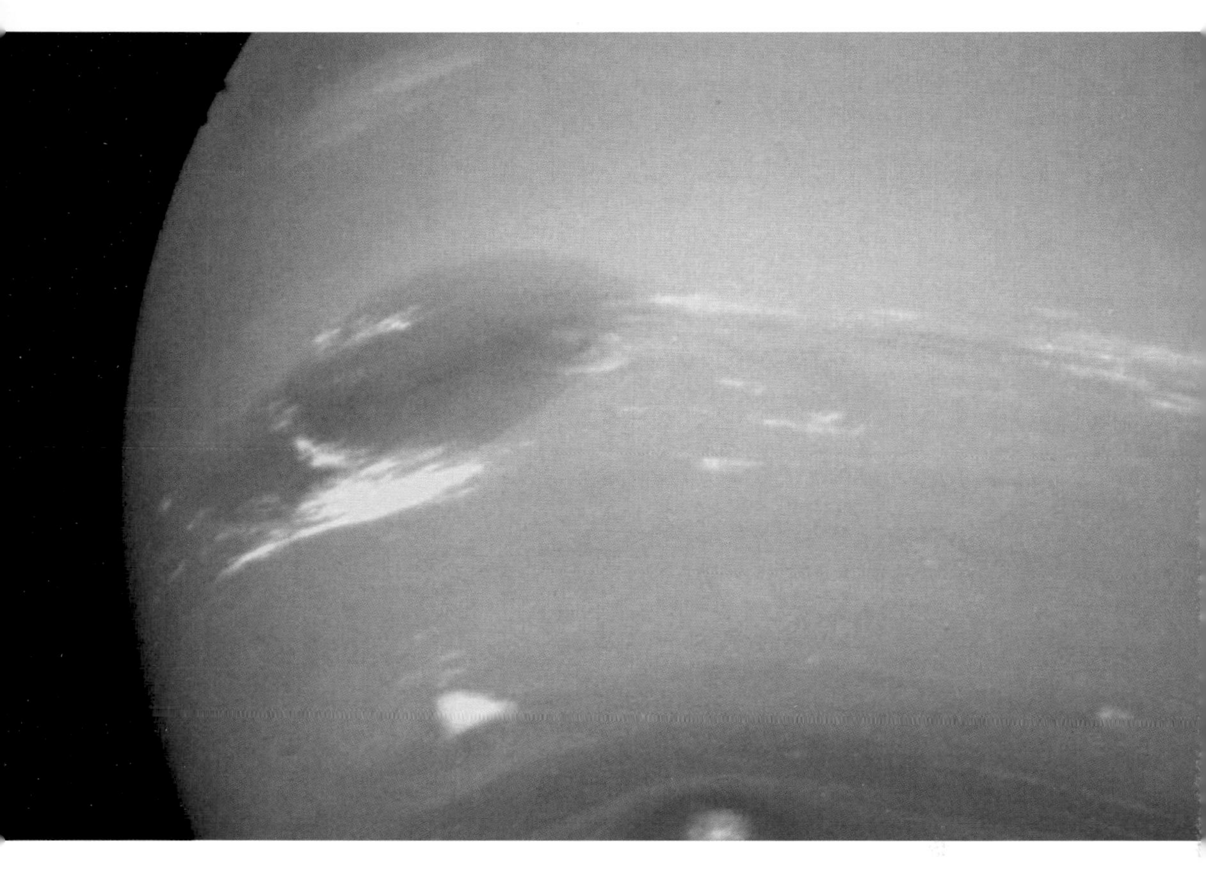

Cloud formations in the latitudinal bands: 27°N (the location of bright streaks), 15° to 25°S (the latitudes of the GDS), 32° to 46°S (surrounding the latitude of 'Scooter') and 68° and 72° (South Polar Features).

Neptune's dynamic climate has some of the strongest winds in the solar system. Most sweep westwards in the opposite direction to the rotation, where wind speeds can reach 2,200 km/hr. The large-scale atmospheric features in the equatorial region have a westward motion of 1,200 km/hr, with smaller features having twice that speed. The winds travel in an easterly direction at high northern and southern latitudes. Once the scientists could image features in the atmosphere, they were able to measure the rotation period of the cloud layer. Cloud features such as the Great Dark Spot were used as a reference frame to determine the rotation period. The results were variable, with the Great Dark Spot giving a rotation period of 18.28 and 18.38 hours. The bright cloud called 'the scooter' at a latitude of 42°S had a period of 16.75 hours, while Dark Spot 2 at a latitude of 55°S gave a rotation period of 15.75 hours. The difficulty in measuring this value is in part due to the distances involved and the ever-changing features on the surface. Voyager 2 also measured the rate of the interior rotation through radio emissions and this was shown to be 16.11 hours. Under all the layers of cloud, a day on Neptune lasts sixteen hours and eight minutes, which means there are no less than 89,666 days in a Neptunian year.

Magnetosphere

When Voyager 2 detected radio emissions from Neptune eight days before its closest approach, this was the first direct evidence for the magnetic field surrounding the planet. The spacecraft had crossed the bow shock at 872,000 km from the planet, thus moving from the region dominated by the solar wind into the planet's magnetosphere. The edge of Neptune's magnetosphere was unusually diffuse because of the highly tilted magnetic field. The magnetic tail extends from the planet millions of kilometres into space.

Neptune has an axis of rotation which is similar to Earth, 29° from the perpendicular, in contrast to the 98° of Uranus. Surprisingly, the two planets are remarkably similar magnetically, with Uranus's magnetic field tilted by some 60° to its axis of rotation, and Neptune's by 47°. Ed Stone, the Voyager project scientist, explains in an interview in 2014 the surprise that this caused:

> One big surprise was a magnetic field. Earth's magnetic field has its pole near Earth's rotational pole, and that's great for compasses. Saturn's the same way. Jupiter's the same way. The small magnetic field on Mercury is the same way. So there was a model that said a magnetic field comes from the rotation of the ionized material in the planet. This was reasonable.

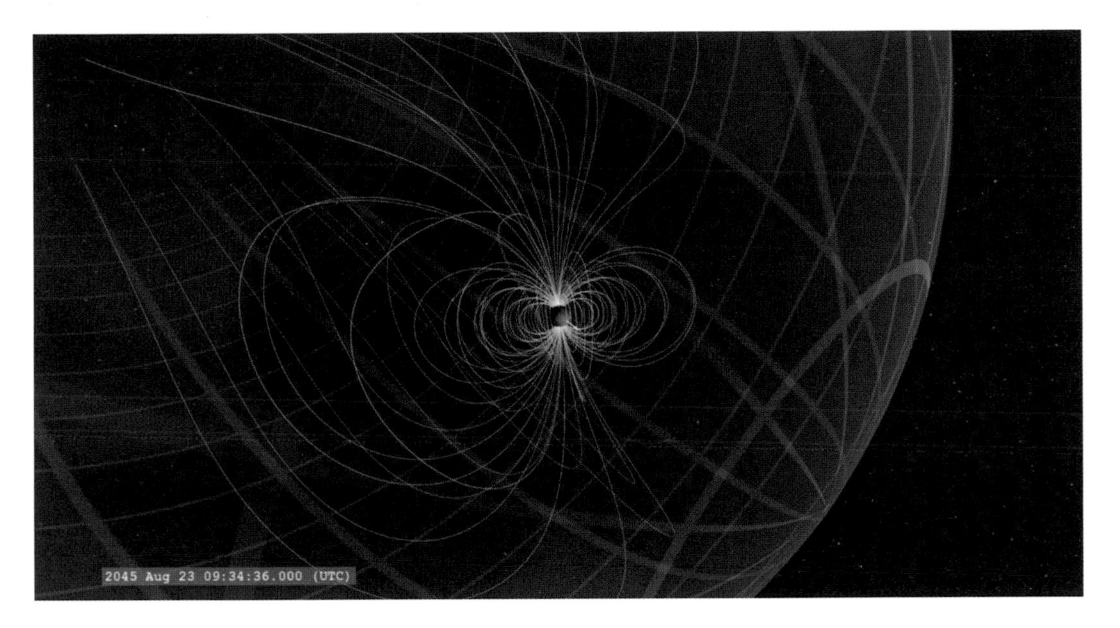

The magnetosphere around Neptune.

> Yet when we flew by Uranus, we found a magnetic pole down closer to the equator than to the rotation pole. The same thing turned out to be true at Neptune as well. So the global current system generating such an offset tilted field was quite distinctly different from what we had believed to be the case for planets in general. We found we had to rethink the geometry of these fluid flows inside planets because we had evidence that they are not always nicely aligned with the rotation of the planet itself.[8]

The magnetic fields of both planets wobble wildly as they spin; they are both 'oblique rotators'. They generate their magnetic fields through convection of electrically conducting material within a thin near-surface shell rather than in a molten core as found on Earth. This wobble was observed by the craft as it moved through the highly charged region of space surrounding the planet.

The dipole tilt and offset of the two ice giants are similar. All the evidence suggests that Neptune's magnetic field is more complex and it could have a quadrupole structure where two dipoles sit next to each other, alternating north then south. The field strength varies across the planet; it is similar to that found on Uranus and at its strongest the field strength is 27 times greater at Neptune than on Earth. This could be a consequence of Neptune's hot interior, but one possible explanation for the complexity of Neptune's magnetic field is that Voyager 2 visited the planet at a period between different magnetic epochs, when the magnetic field is changing its polarity. This reverse in polarity has also been suggested for Uranus and it seems a strange coincidence that a change in polarity would be happening at both planets at the same time.

The plasma contained in the magnetosphere is made of equal numbers of electrons, hydrogen ions and a smaller number of nitrogen ions. The rings and all the moons, except Nereid, pass through this highly charged region at some point during their orbit. Moving through a highly charged and hostile environment affects their chemical composition. In the case of the largest satellite, Triton, it is possible that it contributes to the plasma by injecting heavier ions, particularly nitrogen ions from its geysers, and the more abundant hydrogen ions probably come from Neptune's atmosphere. The moons have some of the darkest surfaces in the solar system and this may be a result of the constant bombardment of the moons by charged particles.

It is thought that lightning storms are frequent events, discharging many charged particles within the upper atmosphere, and although this has been observed on Jupiter, it has not been observed on Neptune. The strikes would not be as dramatic as that seen on Jupiter, which has apocalyptic lightning storms. The electric discharge of lightning over Neptune would have a similar power to that on Earth, because of the ammonia in the atmosphere, which dampens the strength of the lightning strikes. As the lightning storms are weaker, it is unlikely that Neptune's lightning interacts with its rings, a phenomenon that has been seen on Saturn and Uranus. Energetic particles cascading down the magnetic field poles and interacting with the atmosphere create auroral displays on Neptune. Thanks to the tilt in the rotational axis of the magnetosphere, these aurorae can happen across the planet. As future explorers of the solar system, we may find it exciting to stand on the surface of its largest satellite, Triton, and observe Neptune's aurorae from there.

Rings

Voyager 2 was able to confirm the presence of the tenuous ring system around the planet, which had possibly been discovered five years earlier at the European Southern Observatory's (ESO) La Silla Observatory in Chile. The discovery of the rings has been discussed in Chapter Five, but it is worth adding that the Earth-based observations using stellar occultations had only proved successful in 10 per cent of cases and at times ring material had only been observed on one side of the planet and not the other.[9] Hence the flypast gave astronomers their first opportunity to see the rings in detail and confirm their existence once and for all.

The spacecraft would image six rings, which are delicate, especially in comparison to the rings around Saturn, and made of particularly dark material, most likely organic compounds called tholins. In order from the planet they have been named Galle, Le Verrier, Lassell, Arago, a very faint and as yet unnamed ring and Adams. There are variations between the rings, with Galle and Lassell being broad and less distinct. Lassell is the broadest and is optically fainter than all the others. Its material is contained between the edge of the Le Verrier ring and the edge of the Arago ring. The three well-defined rings – Le Verrier, Arago and Adams – have widths of under 100 km with comparatively small depths, making them appear exceptionally slender. Le Verrier has the greatest depth of just 0.7 km.[10] The Adams ring, the outmost ring found at a radius of 62,933 km, was particularly interesting to the

Two bright outer rings are clearly visible around Neptune, along with an inner faint ring at 42,000 km and a faint band extending from 53,000 km to a mid-point between the two bright outer rings. Image taken on 26 August 1989 from a distance of 280,000 km.

scientists. An as yet unnamed partial ring which consists of dust and is very faint is located 1,000 km nearer the planet than the Adams ring. As all the rings are named to honour scientists involved in the discovery of the planet, this ring may provide an opportunity to immortalize Galle's assistant, d'Arrest, in the skies; after all, he suggested using the map that ultimately helped Galle discover the planet. A small moon, Galatea, orbits in the same region as the unnamed ring. It has a role to play, as its orbit causes disruptions to the ring system and arcs found in the nearby Adams ring.

Along the main Adams ring are five regions that have a higher density. Known as 'arcs', they are areas of thicker material and it is possible they were observed during the Earth-based stellar occultation events that produced the original confusing results. Carolyn Porco, who worked on the Voyager imaging team and successfully explained the behaviour of ringlets within the rings of Saturn, has described how 'A fortuitous stellar occultation by the arcs observed by the Voyager ultraviolet spectrometer (UVS) and

On 23 August 1989 the wide-angle camera identified three arcs, which were named (shown here from left to right), Fraternité, Égalité and Liberté (the outer ring is the Adams ring, the inner imaged is Le Verrier's).

photopolarimeter (PPS) experiments . . . confirmed the 15 km arc radial widths observed from the ground.'[11]

The arcs are so unusual that their observation has come under much scrutiny. They have been compared to the only dusty arc in Saturn's G ring, which is held in place with a resonance with the shepherd moon Mimas. Neptune's arcs also have a shepherd moon, the small moon, Galatea, which orbits in the unnamed ring and has an orbit which is in 42:43 resonance with the Adams ring. This moon's resonance was originally suggested as having a role to play in the creation of the arcs; as the moon orbits the planet, it also undergoes a rocking motion with a vertical amplitude of about 30 km as it orbits above and then below the orbital plane of the ring. This allows denser areas on the ring to form where material is shepherded together.

Later research put this hypothesis in doubt and the arcs remain a puzzle. In 2002 Fathi Namouni and Porco proposed that the small eccentricity of Galatea at 0.00022 acted as a necessary driver for the creation and retention of the arcs within the ring. Otherwise the arcs would have no longevity and would disappear; the stability of the arcs is tenuous.[12] Their brightness levels have wavered and there have been variations measured since their discovery. They have been named Fraternité, Égalité 1, Égalité 2, Liberté and Courage; the first three names stem from the motto of the French Revolution and were suggested by the scientists who made their discovery.[13] They are located in a small range of longitude along the ring itself, being found between 247 and 280°. Only a degree or two separate each arc and the smallest two arcs, Égalité 2 and Courage, only extend a degree in size along the ring. Fraternité is the brightest arc and Courage is the dimmest.[14] In 2003 it was reported that the Courage ring had jumped almost 8° along the ring, perhaps skipping over a whole resonance position before settling in a new place. By this point Liberté had nearly vanished altogether and it is unknown if it will re-form.

Location of the Arcs along the Adams Ring		
Name of Arc	Location along Adams Ring	Size in Degrees
(Fraternité), 10	247–57°	10
(Égalité 1), 3	261–4°	3
(Égalité 2), 1	265–6°	1
(Liberté), 4	276–80°	4
(Courage), 1	284.5–85.5°	1

Farewell to Voyager 2

Voyager 2's mission to the outmost known planet in the solar system ended just a few days later. Mission control had been sharing their discoveries with the public as soon as they received them. On the last night the team members had virtually no sleep because they were hosting a programme called Voyager All Night. U.S. Vice

Parting shot of Neptune and Triton.

President Dan Quayle was in the audience and Chuck Berry provided the musical entertainment at the party afterwards. This was 'the last picture show' of the Grand Tour of the solar system by NASA's Voyager missions. The wonders of Neptune had finally been revealed and Voyager 2 had completed its main mission, having achieved far more than the original science team could have imagined. Ed Stone spoke about the highlights of Neptune, saying, 'We knew we would be surprised, we just didn't know how. Neptune had a Great Dark Spot nobody expected, which has since disappeared. And as I said, the magnetic field is also tipped.'[15]

Voyager took a farewell image of Neptune and its largest satellite, Triton. They were both beautiful crescents, the dark side of the bodies dominant as the craft moved beyond them, placing them in shadow while looking back towards the Sun. Gustav Holst's orchestral suite *The Planets* included a quiet fading away of the music in the final movement, 'Neptune the Mystic'. Similarly, Voyager 2 was to fade out of public view for a number of years. The team were still conducting science using the spacecraft. Observations were made of Nova Cygni 1992 in 1992, and there was a failed attempt to see the surface damage produced by Comet Shoemaker–Levy 9 as it impacted with Jupiter in July 1994. Although the craft had a good line of sight, it had moved too far away from Jupiter to detect anything in either the radio or ultraviolet spectrum. In 2007 Voyager 2 moved into a region of the solar system called the heliosphere, passing through the termination shock (where the solar winds are no longer supersonic, because of its interaction with the interstellar medium). Voyager 2 is still travelling and transmitting data. In December 2018 the Voyager team announced that the craft had entered interstellar space in the previous month, a region where cosmic rays were hitting the detectors at increasing rates and the spacecraft was no longer detecting the effects of the solar wind. It had reached a distance of 122 AU (1.83×10^{10} km) from the Sun, over four times the distance of Neptune, and was moving through interstellar plasma, returning readings of temperature and density back to Earth. Voyager 1 had previously moved beyond the heliosphere in 2012.[16]

SEVEN

Neptune after Voyager

Voyager 2's flypast of Neptune gave the first close-up views and tantalizing new information about the eighth planet in our solar system. Since 1989 it has been the task of ground-based and Earth-orbiting telescopes to scrutinize the planet. The work of astronomers using the Hubble Space Telescope, the Keck telescopes and other large telescopes has revealed ever-changing views of Neptune's atmosphere and a deeper understanding of the planet.

Temperature and Internal Heat

The effective temperature of Neptune was something that had proved difficult to measure, with William M. Sinton (1925–2004) of the Lowell Observatory stating, in 1955, with 'present techniques, Neptune and Pluto are probably too cold and remote for study'. In 1977 Robert F. Loewenstein, Doyal A. Harper and Harvey Moseley derived a value for Neptune's temperature using the Kuiper Airborne Observatory's 91-centimetre telescope in far-infrared pass bands 45μm and 93μm. The observations were made over two separate flights and the temperature recorded was 55.5 K. This is still considered the temperature for Neptune at 0.1 bar. The temperature was only slightly colder than the 58 K that had been measured for Uranus earlier in 1977. Uranus is 1.6 million km closer to the Sun and there is considerably less sunlight reaching

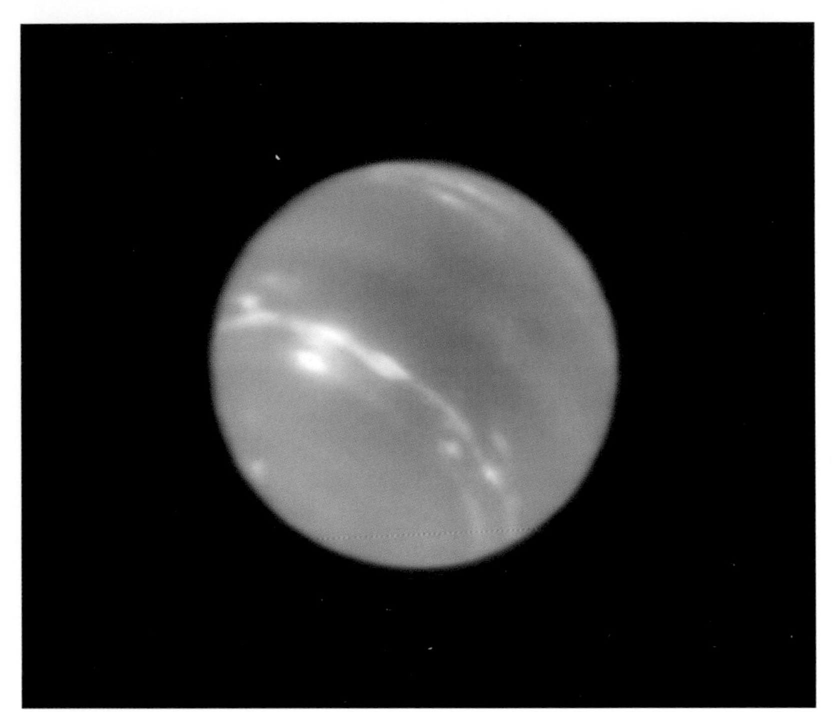

A sharp image of Neptune imaged by the VLT using the MUSE instrument and GALACSI (MUSE's ground layer adaptive optics system), 18 July 2018.

the surface of Neptune. Neptune radiates as much heat from its surface as Uranus and twice the amount of heat it absorbs from the Sun. The source of this extra heat must come from internal compression, and Joshua Tollefson of the University of California, Berkeley, explains that the process is 'largely due to gravitational contraction. As the planet slowly contracts gravitationally, the material falling inward changes its potential energy into thermal energy, which is then released upwards out of the planet.' He suggests that models of Uranus that contain an internal heat source need revisiting, as these temperature measurements confirm that only Neptune has a significant internal heat source, in common with Jupiter and Saturn.

American astronomers William B. Hubbard and Joseph J. MacFarlane refined Ramsey's model (see Chapter Five) and suggested that the three-layer structure enveloped a small rocky inner core consisting of silicates and nickel-iron, making up 25 per cent of the planet as a whole. The pressures in the core are very high, with temperatures up to 5,400 K, similar to the surface of the Sun.[1] Surrounding this hot core is a mantle of hot, dense fluid which consists of water, methane and ammonia. It is confusingly referred to as being icy, but it is a hot boiling mass which constitutes up to 70 per cent of the planet. Deep within this layer the high pressures and temperatures mean that the molecules of water, ammonia and methane begin to break up into their component elements, namely hydrogen, carbon, nitrogen and oxygen. How all these molecules dissociate and mix is little understood. It takes great pressure (a depth of at least 10,000 km) for methane to dissociate into its elements, hydrogen and carbon. In 1981 Marvin Ross suggested that solid diamonds could float within the hot liquid sea and at the deepest layers it may even rain diamonds.[2] This hot, dense interior ocean does not evaporate, thanks to the dense pressure of the atmosphere above it. It is locked in by the cloud layers of hydrogen,

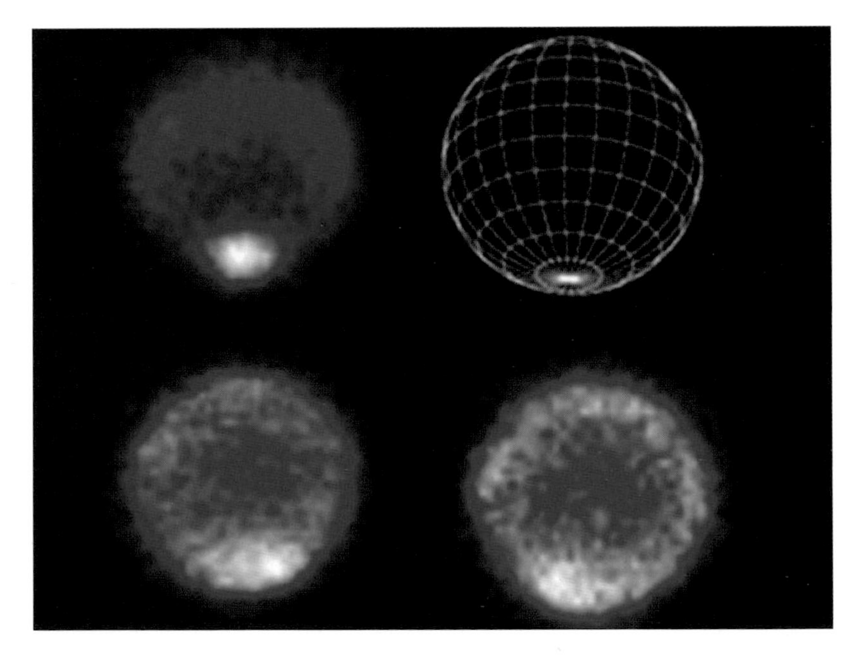

Very Large Telescope images of the interior of Neptune and its radiating heat, with the south pole being hotter than the rest of the planet. The top disc is the troposphere and the bottom two are the stratosphere, taken 6 hours 20 mins apart.

helium and methane, which extend to about 15 per cent of the depth of the planet. The planet contains twice the mass of Earth in hydrogen, which is comparatively small relative to the masses of water, methane and ammonia contained in its mantle. This is the reason the planet is termed an ice giant.

Thermal images of Neptune taken by the European Southern Observatory's Very Large Telescope using the VLT Imager and Spectrometer for the mid-infrared VISIR instrument (VLT Imager and Spectrometer for mid-Infrared) in 2007 revealed that the south pole of Neptune was a small 10 K hotter than the rest of the planet. The telescope was able to sample the temperature in the troposphere and the higher atmospheric region of the stratosphere, where two images were taken 6 hours 20 minutes apart. There is a general trend of warmer temperatures around the south pole, and there is also a hot spot rotating across the stratosphere's southern polar region. Scientists believe the warmer south pole provides a region from which methane can escape from deep within the atmosphere.

Monitoring a Planet from Afar

An early post-Voyager 2 insight into the longer-term weather patterns on Neptune came in 1999. Lawrence A. Sromovsky of the University of Wisconsin–Madison led a team which combined images from the Hubble Space Telescope and NASA's Infrared Telescope Facility on Mauna Kea, Hawaii. He said that 'the character of Neptune is different from what it was at the time of Voyager. The planet seems stable, yet different.' Sromovsky created a time-lapse video of the storms on the planet over the previous few years and the ebb and flow of the weather patterns revealed 'a strange menagerie of variable, discrete cloud features and zonal bands'. There were three distinct weather bands encircling the planet, one which ran around Neptune's equator and two others around the poles. Their continued stability suggests they are as stable as the wind patterns found in Earth's equatorial regions.

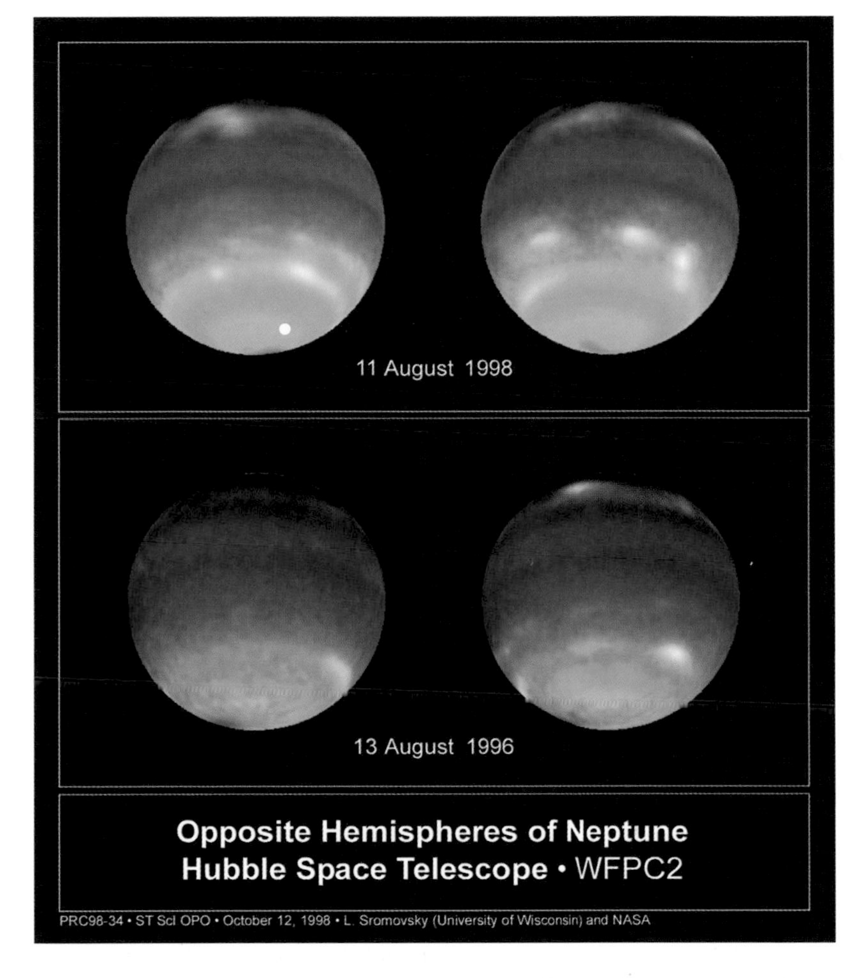

Neptune's wild weather captured in a set of images from 1996 and 1998. This shows the ebb and flow of the blustery winds and moving cloud. Images are from the Hubble Space Telescope and NASA's Infrared Telescope Facility on Mauna Kea, Hawaii.

They also mapped the high-level clouds and found that they varied in altitude. It would take continued observations to draw further conclusions. The team had an opportunity to look for more dark spots as the planet is monitored by the Outer Planet Atmospheres Legacy (OPAL) project and the Hubble Space Telescope images it each year. When they examined images, they confirmed the previous observation from the Hubble Space Telescope that the Great Dark Spot had been replaced by a smaller dark spot, Dark Spot 2. Sromovsky concluded, 'They behave like storms, and the Great Dark Spot was an exaggerated

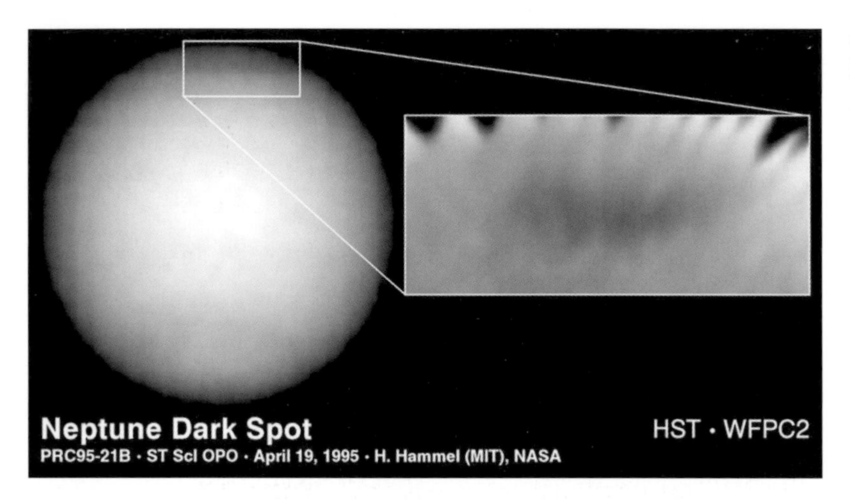

A new dark spot in the northern hemisphere of Neptune, as imaged in April 1995 with Hubble's Wide Field Planetary Camera 2.

feature we haven't seen on any other planet.'[3] The team did not have enough information to know if the intensity of the dark spot would return and if it would grow in size again. The disappearance of the southern hemisphere storms was unexpected. Amy Simon, a planetary scientist at NASA's Goddard Space Flight Center in Greenbelt, Maryland, said, 'We were used to looking at Jupiter's Great Red Spot, which presumably had been there for more than a hundred years.'[4] Questions were raised over the longevity and processes behind the formation of the dark spots. It would take the discovery of more storms in 2015 for some of these questions to be answered.

Storms in 2015 and 2018

In 2015, during the course of the annual monitoring of the planet, a new small dark storm was imaged. Then, three years later, the team discovered there was a large dark spot. 'We were so busy tracking this smaller storm from 2015, that we weren't necessarily expecting to see another big one so soon,' says Simon about the storm, which was similar again to the Great Dark Spot.[5] 'That was a pleasant surprise. Every time we get new images from Hubble, something is different than what we expected.'

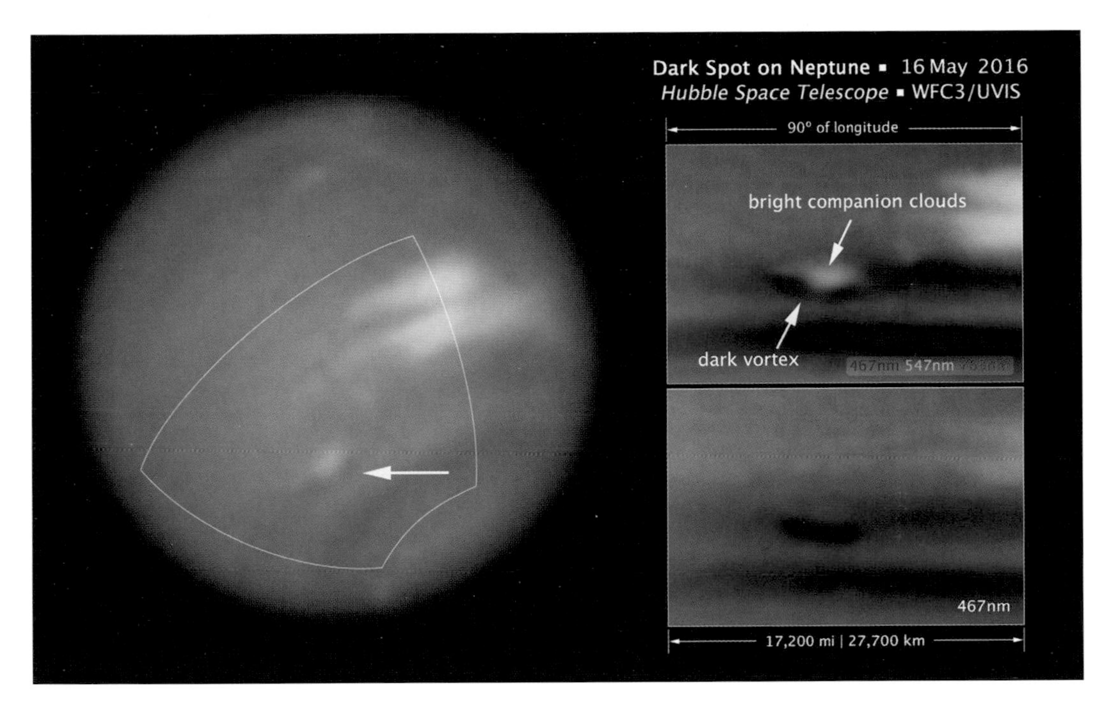

Dark spot on Neptune with its accompanying companion cloud.

Analysing the 2015 image from the Hubble Space Telescope, UC Berkeley research astronomer Michael H. Wong said, 'Dark vortices coast through the atmosphere like huge, lens-shaped gaseous mountains.'[6] It was an observation of bright clouds in 2015 in the region of the later dark spot of 2018 that allowed astronomers to piece together the puzzle behind their formation. The scientists looked through images from 2015 to 2017 and saw white clouds develop in that part of the atmosphere.

Both professional and amateur astronomers on the ground tracked the increase in white cloud activity and found that these white areas were the precursor to the massive dark spot that was observed in 2018. Dark spots are directly linked to these brighter areas. The vortex extends deep into the atmosphere of Neptune, remaining visible on the upper layers of the atmosphere. The 'Berg' on Uranus was thought to have undergone a similar process (see Chapter Three). By contrast the dark vortexes on Neptune have much shorter lifespans. They only exist for a couple of years, from formation to disappearance, and will appear at different latitudes. The decay of one such feature was tracked by the Hubble Space Telescope during 2015 and 2016. Its behaviour was a surprise, as astronomers believed that the vortex would drift (like the Berg) to the equator and break up into smaller clouds. Instead the spot headed towards the pole and faded away. The weather bands around Neptune can explain why this happened. The three broad jets on Neptune are wide and not like those seen on Jupiter, which are tightly constrained by alternating wind directions and create the conditions for the long-lived Great Red Spot. Rather it is thought that the vortexes on Neptune can change 'lanes' and drift between Neptune's westward jet near the equator and eastward ones near the two poles.

Dark spots are estimated to appear on Neptune every four to six years and they will vary in size, shape and drift rates. It is not known how fast these later dark spots rotate. The 2018 spot is roughly 6,800 km in diameter, a major storm which was even larger than the one observed in 2015.[7] A year later it was imaged again and seemed

The demise of a dark spot on Neptune's southern hemisphere. Its drift towards the pole leads to a two-year fading of the vortex, shrinking from 5,000 km across to 3,700 km across.

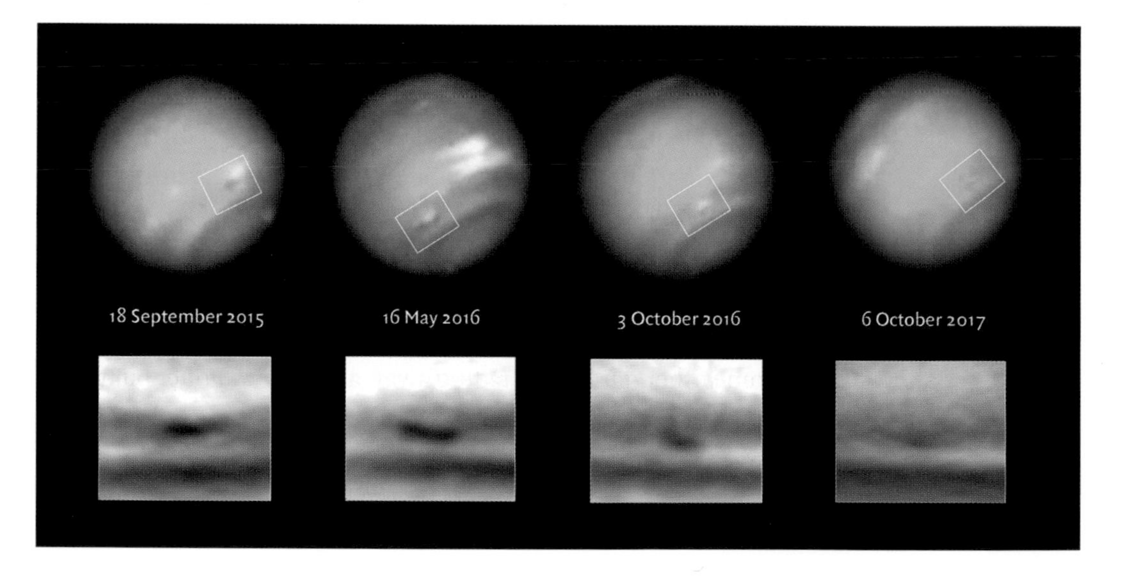

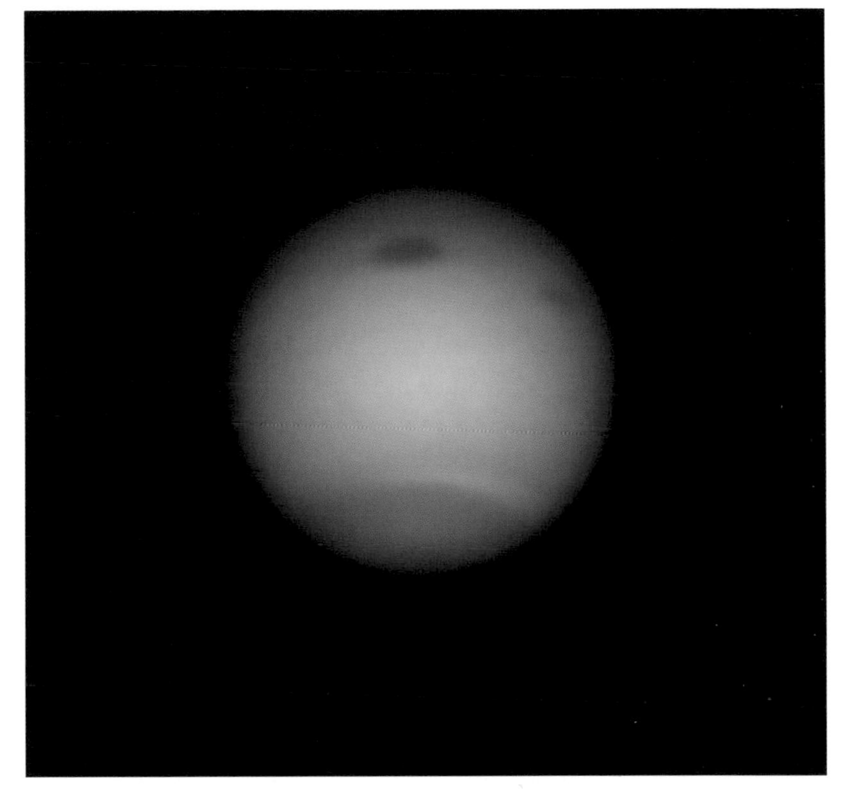

A dark spot the size of the Atlantic Ocean and the emergence of a smaller dark spot. The larger of the two was drifting towards the equator before it made a U-turn in August 2020 and started on a path towards the polar region.

to be migrating towards the equator, where it was thought that it would dissipate. In the latter part of 2020 the 2018 spot started to behave in a manner that was unexpected. It changed direction and then moved away from the equator. In doing so it fragmented, with a large piece of the spot breaking off and then disappearing. 'We are excited about these observations because this smaller dark fragment is potentially part of the dark spot's disruption process,' says Wong. 'This is a process that's never been observed. We have seen some other dark spots fading away, and they're gone, but we've never seen anything disrupt, even though it's predicted in computer simulations.'[8]

A Surprising Equatorial Storm

In 2017 as the Keck Observatory imaged the planet, a vast storm unfolded on the surface, appearing as white scarring on the images. This was surprising, as storms do not normally appear in the equatorial region. This was the first time they had seen this happen and it was a storm about the size of Earth. 'Historically, very bright clouds have occasionally been seen on Neptune, but usually at latitudes closer to the poles, around 15 to 60 degrees north or south,' explained Imke de Pater of the University of California, Berkeley's Astronomy Department. 'Never before has a cloud been seen at, or so close to, the equator, nor has one ever been this bright.'[9] The atmosphere of Neptune undergoes changes with latitude, like the other planets in the solar system. But this storm spanned over many latitudes and it was a mystery as to how this could form. 'This big vortex is sitting in a region where the air, overall, is subsiding rather than rising,' said de Pater.[10] This would be a cooling region, as the gases would rise when hot, condense

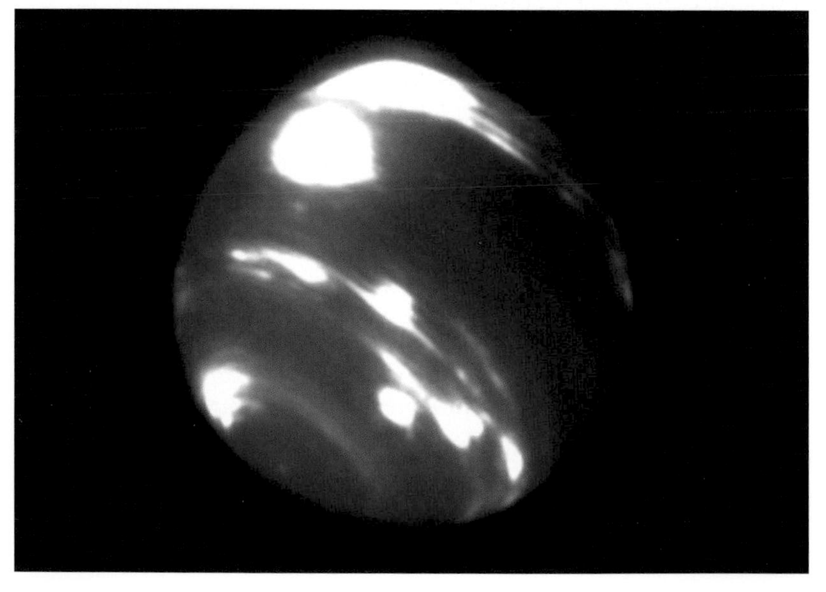

Neptune imaged in 2017 with the Keck Telescope, showing a gigantic storm developing near the equator.

and then form clouds. Overall, the longevity of such a feature is questionable, as weather systems on Neptune such as the Great Dark Spot have come and gone in a relatively short amount of time through a process of disruption which sees them travel across the disc, fragmenting then disappearing. De Pater remarked, 'This shows that there are extremely drastic changes in the dynamics of Neptune's atmosphere, and perhaps this is a seasonal weather event that may happen every few decades or so.'[11]

As measurements are currently confined to the upper layers of the atmosphere, it is unknown how features change at lower altitudes; only a return to Neptune and releasing a probe into the atmosphere would be able to make observations that could take measurements such as wind speed at deeper depths. This is, in part, the reason behind a number of current suggested missions being discussed by NASA and the ESA.

Future Missions

A return to Neptune is considered a high priority, along with its neighbour Uranus, as it has yet to have a dedicated planetary mission. There are a number of reasons that a mission is seen as advantageous; these include further examination of the planet's atmospheres, magnetospheres, heliospheres, rings, moons and the questions surrounding the origin and formation of the planet. Furthermore, as Neptune-class exoplanets make up the largest group of currently known planets, a mission to Neptune would shed light on exoplanets as well as our own ice giants.

A number of missions have been proposed, although some, such as the twin orbiter ODINUS proposed by ESA as part of its Cosmic Vision Campaign (2005–25), have been abandoned. The expense and time it takes to get a craft to this outer region of the solar system has led to a collaborative idea between NASA and the ESA called OSS (Outer Solar System). With instrumentation similar to what was carried on New Horizon, the mission envisages that it could attain and surpass the results that were obtained by Voyager 2. The mission would prioritize the measurement of the gravitational fields out to 50 AU. This is a medium M-Class mission, and the concept was first proposed in 2012. A mission of this size would cost in the region of €500 million. There is the argument that with the expense and extensive time to get a mission to Neptune, an entry probe rather than a cheaper flyby mission would offer more scientific value, and a large-class mission similar to the JUICE (Jupiter Icy Moon Explorer), planned for launch in 2022, would be more appropriate and visionary. A mission called Trident proposed by NASA's Discovery Program in 2019 will be discussed in Chapter Eight, alongside the Triton Hopper concept lander.

Other Neptunes

Before we leave the ice giants it is worth visiting similar planets that are being discovered around other stars. Exoplanets were first discovered in the 1990s and the first to be identified as a Neptune-like planet was observed in 2014, within the Milky Way in the direction of Sagittarius. By the end of 2020, there were more than 4,300 confirmed exoplanets, with nearly 1,500 of these being Neptune-like. These exoplanets are a similar size to Uranus and Neptune and are thought to have a similar composition to the ice giants, being composed mainly of hydrogen and helium and some water, ammonia and methane. The atmospheric composition of the exoplanets can be gleaned from information collected by the Spitzer Space Telescope, Hubble Space Telescope and, most recently, NASA's Transiting Exoplanet Survey Satellite (TESS). The light from the parent star passes through the atmosphere of the planet and, using spectroscopy, astronomers can detect the molecules in the atmosphere of the planet. Neptune-sized planets have thick atmospheres which can make it difficult to detect molecules. By good fortune, a Neptune-sized planet has been found that was relatively cloud-free, thus enabling measurements to be taken. In 2017 the atmospheric signature of HAT-P-26b was measured. This is a 'warm' Neptune, as it orbits closer to its star than Neptune orbits our Sun. The system is 400 light years distant from us and its atmospheric signature was a surprise. Composed almost entirely of hydrogen and helium, it has a primitive atmosphere in comparison to Neptune. The atmosphere had a strong signal for water, but it is felt that HAT-P-26b is not a water world and this signal is difficult to understand. 'Astronomers have just begun to investigate the atmospheres of these distant Neptune-mass planets, and almost right away, we found an example that goes against the trend in our solar system,' said Hannah Wakeford, a researcher at the Goddard Space Flight Center, Maryland.[12]

Gliese 436b was discovered in 2004 by R. Paul Butler and Geoffrey Marcy of the Carnegie Institute of Washington and University of California, Berkeley, using the radial velocity method, a technique which exploits the 'wobble' produced by the gravitational tug of the planet on the parent star. By 2007 continued transit observations of this planet enabled astronomers to determine its

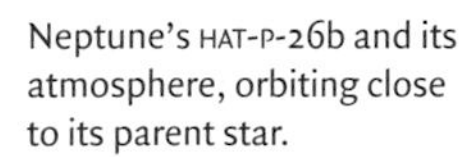

Neptune's HAT-P-26b and its atmosphere, orbiting close to its parent star.

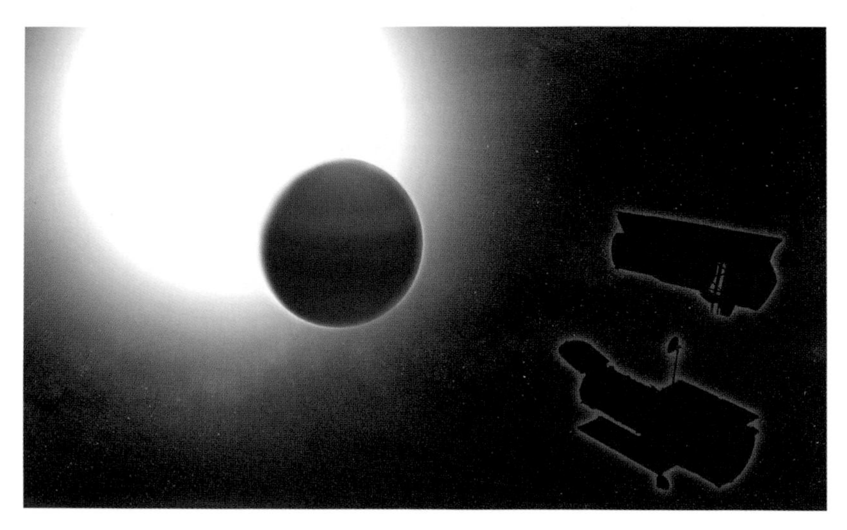

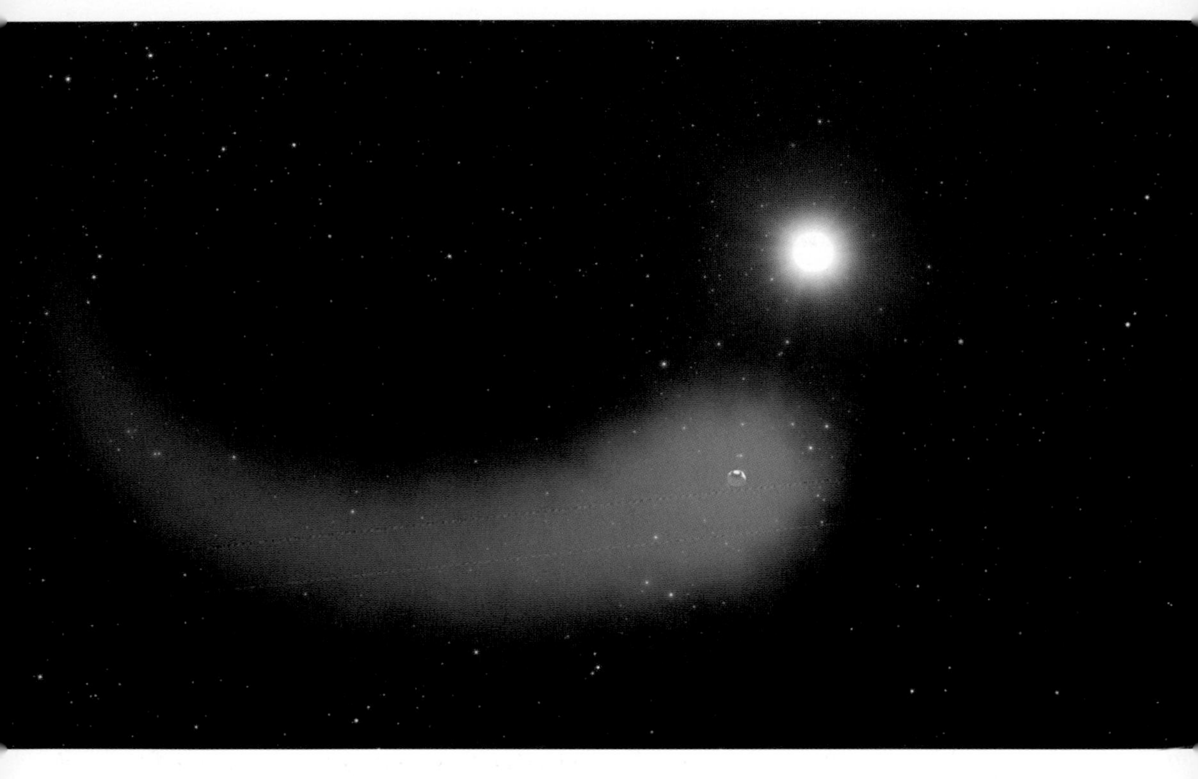

This artist's impression of Gliese 436b shows the large comet-like tail that is being burnt off from the planet as it orbits close to its parent star.

mass and orbit. Similar to Neptune in size, this planet has a semi-major axis of just 0.028 AU, and it orbits fifteen times closer to its parent star than Mercury does to our Sun. The very high temperatures of 712 K have given this type of planet the name 'hot Neptune'. These are planets that are Neptune-like in size and are thought to migrate towards the parent star. Like the so-called 'hot Jupiters', this new type of planet has confounded scientists since it was first recognized. Its surface temperature is estimated to be higher than if it was only heated by stellar radiation. An exceptionally high-pressure interior suggests that water is compressed into a solid state that is denser than ice, and the temperature of the planet is thought to be higher than measurements suggest. There is abundant atmospheric carbon, although methane is lacking and the infrared Spritzer telescope spectrum of the planet suggests it is predominantly helium.[13]

This bizarre world is currently orbiting in less than two days; its atmosphere is evaporating, leaving a trail of hydrogen, producing a gas cloud and a long comet-like tail. It is expected to have lost 10 per cent of its mass in the first billion years of existence and has a lifespan that would take it beyond the life of its parent star.[14]

The first 'ultra-hot Neptune', LTT 9779 b, was observed by TESS in 2020. It has an extremely short orbital period of just nineteen hours.[15] It is orbiting within what is called the Neptune desert, a region around a star where very few Neptune-sized planets have been found. Being so close to the parent star, its surface temperature is well over 1,700 K. The unfortunate planet's outer layers are starting to be stripped and transferred to its parent star. This demolition is probably the reason why these planets are rare.

Exoplanets plotted by their size and distance from their star (in miles). A region called the Neptune desert can be seen, where no Neptune-sized planets orbit close to their star. The exoplanet plotted, shown by a red dot, is on the edge of the desert region and is undergoing changes due to the evaporation of its atmosphere.

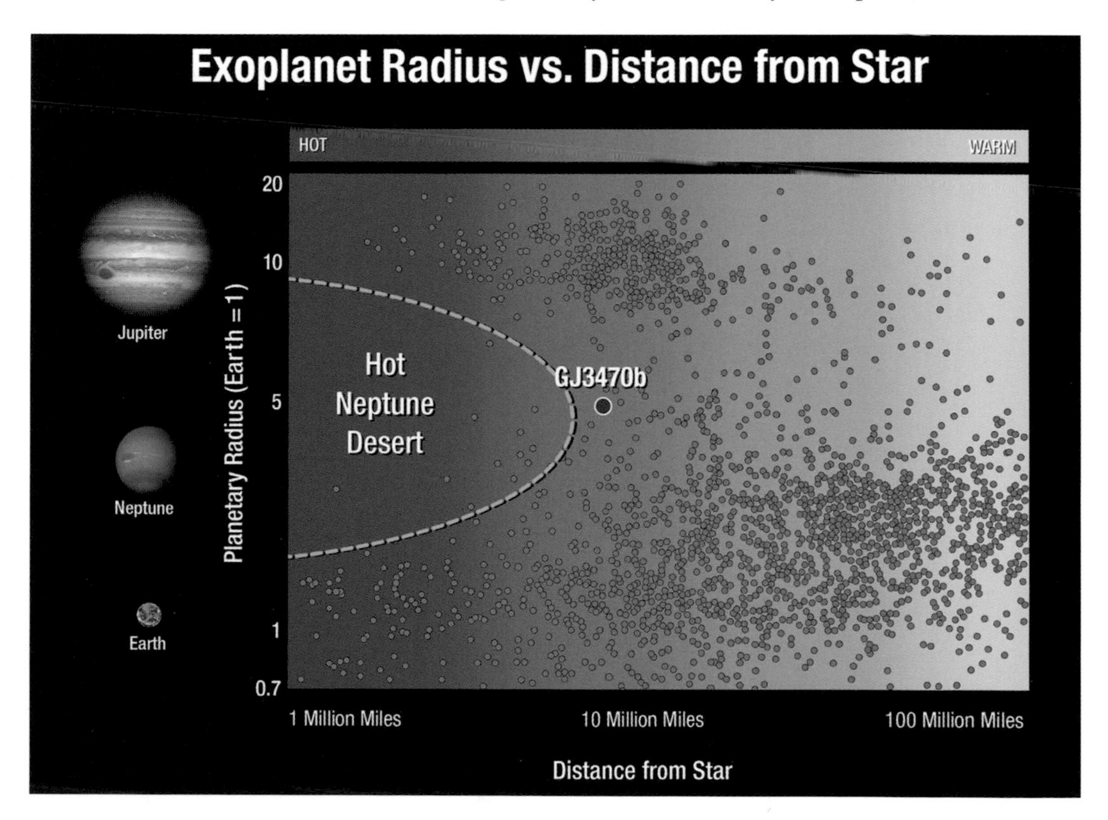

Another category of exoplanets that could resemble our ice giants are mini-Neptunes; these are also known as 'gas dwarfs' or 'super Earths'. Less massive than Neptune, but larger than Earth, this classification of planet is defined by the planets' mass and not by their composition, temperature or orbital characteristics. Some of these planetary bodies may resemble Neptune, as they have thick atmospheres made of hydrogen and helium and icy layers of rock. Others are considered to be super Earths. These are probably very bloated rocky planets, with an ultra-thick atmosphere made from water, or a possible liquid ocean consisting of water, ammonia or heavier volatiles.[16] These planets are currently the most common exoplanet discovered, although we do not have any examples of them in our solar system. As more exoplanets are observed, what was thought to be improbable is becoming commonplace. There is much to be learnt from exoplanets. Their location, mass, variability, evolution and chemical composition can teach us more about Neptune and Uranus. The results arising from the study of Neptune-like planets much further afield than our own solar system are key to understanding our own ice giants.

EIGHT

Moons of Neptune

Triton dominates the other thirteen known satellites of Neptune; it is the largest moon of Neptune and the sixteenth largest object in the solar system. Looking up at Triton from Neptune, it would appear similar in size in the sky as our Moon does from Earth. Triton's orbit is highly inclined to the planet's equator at 157° relative to the orbital plane and it is retrograde – the orbital motion of the moon is opposite to the planets. It is therefore suggested that this unusual motion demonstrates that Triton did not form within the solar nebula with Neptune itself. If Triton is a captured object, it was ensnared early within the formation of the system and it tore through the existing Neptunian moons, thus restructuring the system. While doing this, it cleared the Neptune system of all other large bodies and consists of 99.5 per cent of the total mass of the moons. Seven of the moons are regular moons and thought to have formed from the accretion disc. They have a prograde motion. The remainder are irregular moons, captured from the nearby Kuiper Belt or resulting from debris created by Triton's destruction of the original system. They follow orbits which are found far away from the planet itself. The moons of Neptune are all named after Greek and Roman water gods.

Triton

Triton was discovered in 1846 just seventeen days after Neptune itself. William Lassell had turned his scope to the newly discovered planet, and after eight nights of observations he was to observe this large moon. Lassell was a distinguished British amateur astronomer who made his fortune in the brewery trade. He used his self-built 24-inch-aperture metal mirror reflector on an equatorial mount at his observatory in West Derby, Liverpool, to observe the Neptunian moon. The object was only 14th magnitude. By 1847 Lassell had made sufficient observations to determine the moon's orbital period and its distance from the planet:

> Unfavourable weather, and the low altitude of the planet, have not allowed Mr. Lassell to observe the *ring* of *Neptune* satisfactorily; there is, however, no doubt in his mind as to the existence of a ring. The observations of the satellite have been more successful: it has been seen repeatedly in the course of the year, and the non-existence of any star in the places successively occupied by it, frequently ascertained. From the mean of his observations, Mr. Lassell concludes that the satellite revolves about the planet in 5d 21h nearly, and that its greatest elongation is somewhere about 18″.[1]

This is remarkably close to the current-day values of 5 days 21 hours and 3 minutes and an elongation of 18 seconds of arc. It is now known that Triton takes a leisurely 5.8 Earth days to turn once on its axis; this is the same time in which it makes one orbit of the planet. It is tidally locked within a synchronous orbit, with one face always orientated in towards the planet, and it orbits in retrograde.

The name Triton, after the Greek sea god who was the equivalent of the Roman god Neptune, was proposed in 1880 by the French astronomer Camille Flammarion (1842–1925) in his book

Astronomie Populaire. It remained unofficial but extensively used in the late Victorian period and early twentieth century. The American astronomer Henry Norris Russell (1877–1957) proclaimed in 1932 that the name 'might as well achieve general recognition'.[2]

Pluto was discovered by Clyde Tombaugh (1906–1997) in 1930. For most of the time, it orbits within the Kuiper Belt, outside Neptune's orbit, taking 248 Earth years to complete one orbit. For thirty of these years it crosses inside the orbit of Neptune and is closer to the Sun; it is unlikely to ever collide with the planet. Its similarities to Triton are quite remarkable. Physically, they have a similar density, size and surface temperature, which suggests that Triton formed in the same region as Pluto and is a body captured from the Kuiper Belt. In 1936 Raymond A. Lyttleton (1911–1995) suggested that Pluto may be an escaped moon of Neptune, although this is unlikely and it has most likely formed from material in the Kuiper Belt.[3] Detailed observations were made of Triton in the 1930s after Pluto's discovery. The mass of the moon proved difficult to measure and George Ellery Hale (1868–1938), the founder of the Mount Wilson Observatory, wrote that measurements made during 1931 gave a poor result, with a moon mass of between one-tenth and four one-hundredths that of Earth's.[4]

A Second Moon – Nereid

After the discovery of Triton it would be over a hundred years before a second satellite was observed. Astronomers had been looking for moons around Neptune; none were observed until Gerard Kuiper observed Nereid on 1 May 1949. Previously there had been a number of failed attempts to find new satellites. An unconfirmed object was reported by the German-American astronomer John Martin Schaeberle (1853–1924). In 1892, using the Lick Observatory telescope, he spotted a faint moving object just 24 arc seconds from the planet. A search in 1929 for further satellites made by William H.

Christie (1897–1955) using the 60-inch telescope at Mount Wilson was fruitless.[5]

Kuiper discovered the new moon in photographic plates taken with the 2-metre telescope at the McDonald Observatory, Texas. At first light in 1939 this telescope was the second largest in the world. A dim 19th-magnitude object, Nereid has a much smaller diameter to Triton. With a diameter of 340 km it is one-eighth of Triton's size. The orbit of Nereid around Neptune takes it far outside that of the orbit of Triton. Its highly eccentric orbit has a mean distance of 5.5 million km from the planet and it takes the moon nearly a whole Earth year to orbit around Neptune. Nereid moves seven times faster at periapsis than when at apoapsis (periapsis being the smallest distance and apoapsis being the largest distance from the planet). Kuiper then turned his attention to Triton. Using the 200-inch Hale reflector at Palomar in 1954, he made measurements of the disc and estimated the moon's diameter as 3,800 km, much larger than the current value. The radius was a contested issue for many years; the estimates ranged from between 2,500 and 6,000 km. Its currently accepted value is 2,705 km. One difficulty with making measurements of the angular diameter of Triton from Earth is that it is highly reflective and has an albedo of 0.719. Early searches for an atmosphere around the moon in 1969 by the American astronomer Hyron Spinrad (1934–2015) failed to produce any results. Then in 1988, a year before Voyager's arrival, Dale P. Cruikshank made near-infrared spectrophotometric observations of the moon and was able to confirm the presence of an atmosphere. The atmosphere was predominantly methane, with a significant percentage of nitrogen.

Harold J. Reitsema, William B. Hubbard, Larry A. Lebofsky and David J. Tholen were making ground-based stellar occultation observations on 25 May 1981 with two University of Arizona telescopes, the 1.5-metre reflector at the Catalina Station and the 1-metre reflector on Mount Lemmon, when there was an 'abrupt occultation event'. The team suggested that Neptune had a third satellite, 180 km in diameter, and an orbital distance some three Neptune radii from the centre of the planet.[6] The moon was designated S/1981 N1 and would not be observed again until the Voyager flypast, when it was designated S/1989 N2 and later named Larissa.

With so few moons known before the Voyager 2 flypast in 1989, the team were keen to have a close flypast to Triton, as it was only one of three bodies known to have nitrogen in its atmosphere. Cruikshank had suggested that Triton's surface would be frozen methane rather than rocky, and Ellis Miner wondered if on arrival they would find a 'strange moon', with nitrogen-filled lakes and methane ice in a 'sludgier' form. With the close approach, Triton started to give up its secrets.

Voyager's Flypast

In May 1988, and over a year before its closest approach of Neptune, Voyager 2 was imaging the moon Triton. Although at a distance of 684 million km, with Triton appearing as a pale red dot, these images were far better than anything obtained hitherto here on Earth.[7] As Voyager approached the moon, it was possible to get an accurate measurement of the diameter for the first time – of 2,700 km – and on 25 August 1989 Voyager 2 would make its closest pass to Triton at a distance of 40,200 km. During the flypast it was learnt that the moon has the coldest-known surface in the solar system, being a frigid 38 K. At this temperature nitrogen condenses onto its surface as a frost layer. As in the case of Pluto, its surface is predominantly made of this element. In addition to nitrogen, the surface is made from a range of low-temperature volatiles including carbon dioxide, carbon monoxide and methane.

The surface has been divided into three distinct regions: Uhlanga Regio is the polar region; Monad Regio, the eastern equatorial; and Bubembe Regio, the western equatorial. Features

Triton, photographed by Voyager 2 in 1989.

on Triton are named after sacred sites and mythological places and creatures. Uhlanga Regio is named after the Zulu reed swamp from which humanity sprang, Monad Regio after the Chinese symbol of duality and Bubembe Regio after Bubembe Island, home to the temple of Mukasa in Uganda.

Monad Regio forms the green-blue smooth area around the equator, on the eastern half of Triton. It shows a relatively new surface of nitrogen ice, being several million years old, which is just a small fraction of the age of the solar system.[8] Despite being the flattest region of the surface, there are still plateaus, sunken lakes, pits and craters. The Monad Regio has unusual dark mushroom-shaped features (guttae); they are very similar in appearance to Martian dark dune spots (also known as 'Dalmatian spots' and 'fried eggs'), although they are smaller in size. Formed with the boundary to the Uhlanga Regio, these are probably due to the long seasons on Triton, resulting from a change in temperature as the moon emerged from a particularly hot and long-lasting spring which would drive change in the consistency of the surface ices.

Ed Stone considers Triton to be one of the highlights of the flypast of Neptune. In a conversation with Francis Reddy, a senior science writer for the Astrophysics Science Division at NASA's

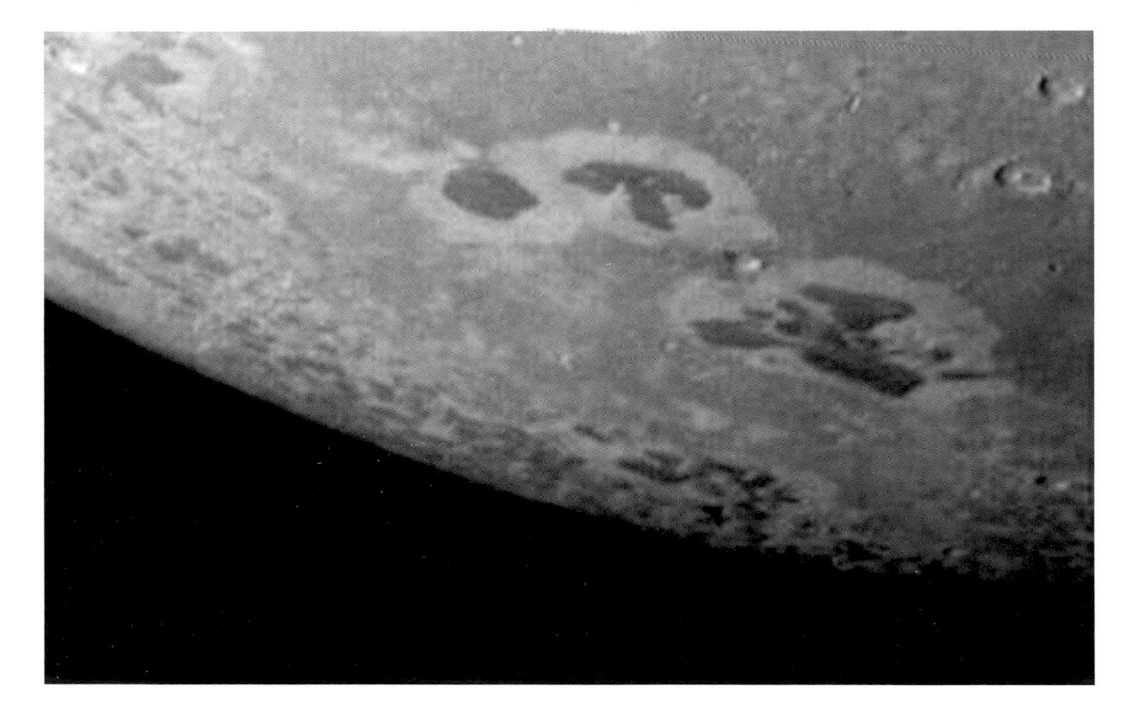

Mushroom-shaped features in the Monad Regio may demonstrate the effects of the long seasons. The largest crater visible on the right is Mazomba.

Goddard Space Flight Center in Greenbelt, Maryland, he recalled the encounter:

> Neptune gave us another big surprise. That was the moon Triton, which is about the same size as Pluto. It probably started out like a Pluto but it was captured by Neptune, presumably into an elliptical orbit. And then as that orbit circularized, all of that orbital energy ended up melting this moon, which has a lot of water ice. We saw a surface unlike any we had seen. Even at very cold temperatures, we saw active geysers erupting from that polar cap of frozen nitrogen.[9]

Triton is geologically active like the Earth, Jupiter's moons (Io and Europa) and Saturn's moons (Enceladus and Titan). As a result of tidal heating, there are volcanos and fractures on the surface of this icy world. It has few impact craters and much of the surface is considered to be young; the dark streaks across its face are thought to be dust deposits from volcanic eruptions. The flypast witnessed a spectacular cryovolcanic eruption in the Uhlanga Regio on 26 August 1989. The image sequence shows plumes of dark material erupting upwards, 8 km into the air. The dark particles drifted for at least 150 km across the landscape. This suggests that a layer of darker material made of organic compounds is present under the surface of pale nitrogen ice. When the sunlight shines on the polar regions the darker material absorbs the sunlight at a faster rate, and it warms the surrounding nitrogen ice, which turns into a gas. The pressure increases as the gas expands and this results in the explosive eruptions. Triton is one of only a few moons with a known atmosphere; Saturn's moon Titan is another. The atmosphere around Triton consists of nitrogen, which is being replenished by cryovolcanic activity. In the images taken by Voyager, dark particles can be seen being carried away downwind in the thin atmosphere and this shows the wind direction at the surface, with much of

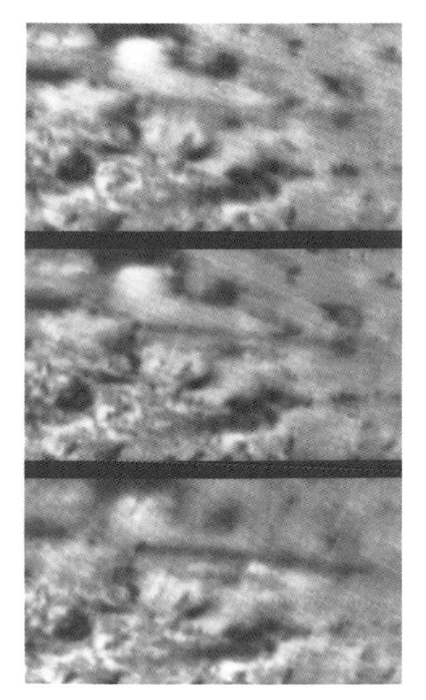

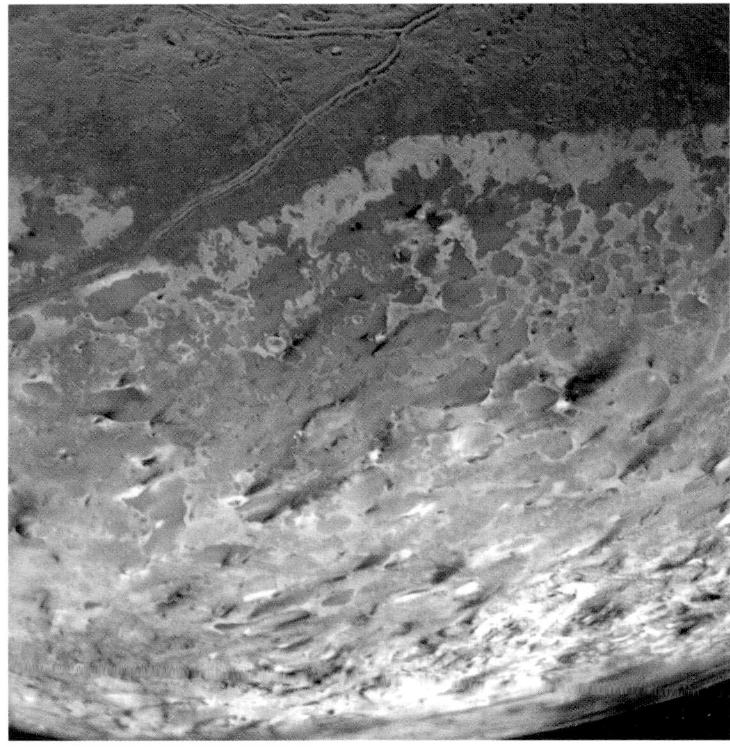

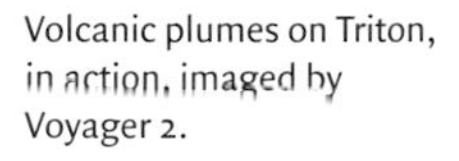

Volcanic plumes on Triton, in action, imaged by Voyager 2.

Namazu Macula is one of many dark streaks across the surface (central right). Resulting from the deposit of particulates, this dark steak extends 75 km.

the motion extending northeast and only a few deposits blowing towards the west. There are currently only a few volcanically active worlds in the solar system. It is unlikely that the volcanic eruption witnessed by the flyby was a rare event and Alan Stern says,

> After Io and Europa, Triton appears to be the most active outer solar system satellite (Titan being an unknown at the moment), and the time-averaged volumetric resurfacing rate on Triton implied by our results is similar to estimates for Venus and the Earth's intra-plate zones. Given the geological evidence that much of the recent volcanic activity on Triton is deep-seated (as opposed to surficial and insolation-driven), a logical source for the lavas is a perched layer of low melting point materials such as aqueous ammonia and/or methanol – an internal ocean.[10]

During the flypast a thin, diffuse haze in Triton's troposphere was imaged at its limbs. This photochemical haze forms when hydrocarbons and methane react with sunlight, much in the way that smog is formed in Los Angeles. The interaction between the atmosphere and surface is, so far, little understood. There are clouds; these can be seen at the limbs and are optically denser patches than the haze. At the surface, the atmosphere has a pressure of just 14 microbars and it is in equilibrium with the ground ice of nitrogen at surface level.[11] Nitrogen condenses and evaporates at the surface and it's one of a few solar system bodies whose main chemical atmospheric component does this (another example being carbon dioxide, a major constituent of the atmosphere of Mars), but this interaction is complicated by factors such as the internal heating of the moon.

The moon itself is going through a process of global warming and temperatures have risen since the Voyager flypast. The atmosphere is rapidly doubling in bulk. In the nine-year period from 1989 to 1998 the American astronomer James Ludlow Elliot and his team showed that this leads to an atmospheric increase in temperature of about 5 per cent, which would equate to a rise of over 11°C here on Earth. 'At least since 1989, Triton has been undergoing a period of global warming. Percentage-wise, it's a very large increase.'[12] Elliot believes this increase in temperature heralds the approach of the warmer summer period at the southern pole, which leads to significant frost migration from one polar ice cap to the other.

The change of the seasons results in a darkening of the surface albedo which leads to increased heating and higher-than-expected seasonal variations in temperature. The moon has a rotational axis of 40° from Neptune's orbital plane, dipping one pole and then the other towards the Sun as it moves through the course of a year, and this gives rise to seasons. Due to the synchronicity between Triton and Neptune, each of the seasons on Triton lasts a long time:

Triton's complex atmosphere, as imaged by Voyager 2.

spring, summer, autumn and winter all last 41 years or a quarter of Neptune's orbital period. The arrival of summer in 2010 brought about seasonal changes, with the southern hemisphere pole warming and a subsequent thickening within the atmosphere.

The polar region named Uhlanga Regio has a pinkish hue, which is produced by the chemical reaction of methane with sunlight. Due to the seasonal changes and the move from spring to summer, the predominantly nitrogen-based water ice is burnt off in the extended but feeble sunlight hours. Its sublimation means that it is being redeposited in the now-dark northern hemisphere.

The western equatorial region, Bubembe Regio, is entirely covered by a type of geology called 'cantaloupe terrain'. There is little cratering on this area, which suggests it is incredibly young.

Bubembe Regio, a cantaloupe region on Triton, where fissures and depressions give it a resemblance to the skin of a cantaloupe melon.

The terrain is scarred with bumps and ridges which are created from rising ice and lava flows composed of water and ices where less dense material rises through a denser subsurface known as 'diapirs'. This process creates a structure that appears well-ordered; it features elliptical and kidney-shaped depressions that are closely spaced and fairly uniform. They can reach up to 40 km in diameter and extend deep into the surface. A similar surface can be seen on Jupiter's moon Europa, where it has been studied in much greater detail. The terrain is criss-crossed with interconnected ridges stretching kilometres across the surface, which is why it is named after the cantaloupe melon. The melting and resurfacing on Triton could have happened during its capture, which would have put the moon under great stress. The heat generated by this event would have kept the surface in a near-liquid state for up to a billion years. Recent research in astrobiology considers these dynamic features to be very exciting, as it suggests there could be the right conditions for microbial life to form. Extremophiles, organisms that thrive in extreme environments (similar to cryophiles found on Earth), may have evolved to cope with the moon's conditions. This strange form of life may exist there in a subsurface liquid ocean. Mathematical models of the interior show Triton has a solid core, mantle and crust similar to Earth; the mantle is made from water, with up to two-thirds of the moon being made of a rocky metal core. Because of its large size, the core would have enough rock within it to power radioactive decay and convection in the mantle, generating heat and allowing for a subsurface ocean to form.

The capture of Triton by Neptune has had an enduring effect on the surrounding system. Triton's addition to the Neptune system would have happened slowly, as its orbit had to lose momentum. The speed of its approach was critical: too fast and Triton could have smashed into the face of the planet, or even been ejected from

Final image of Triton from Voyager 2, taken on 25 August 1989 at a distance of 90,000 km.

the system altogether; its capture must have been slow enough for it to finally reach its current tidally locked orbit with Neptune. Nonetheless, this was still a violent and cataclysmic event for the existing satellites around Neptune. Many of the existing moons are second generation and were formed from the debris of satellites torn apart by the arrival of Triton. One interesting theory to explain Triton's capture involves it being part of a binary system in which only Triton was retained and its twin ejected into the surrounding Kuiper Belt. It is possible but unlikely that this moon was Pluto. However it happened, this event occurred billions of years ago. These satellites are known as second-generation moons, as they did not form with the planet itself. This system is still dynamic, and most recently a third-generation moon has been discovered called Hippocamp. This is a broken-off fragment of a second-generation satellite, possibly from an impact by a comet, which will be discussed in more detail below.

Nereid was imaged during the Voyager 2 flypast, but it was poorly positioned at that time, being over 4.7 million km away from the spacecraft.[13] The pixelated image shown here is an elongated object, which has since been confirmed as being moderately ovoid-shaped by infrared observations from the Spitzer and Herschel space observatories.[14] Kuiper surmised that the object had most likely been captured from the neighbouring Kuiper Belt, in much the same way Triton had been. More recently this has come into question. In 1998 the detection of water ice on the surface of Nereid by Michael E. Brown of the California Institute of Technology showed that the surface was a combination of water ice and a dark material similar to Uranus's satellites Umbriel and Oberon. He suggests that this demonstrates the moon is a regular satellite which formed in situ around

Nereid, imaged by Voyager 2, on 24 August 1989.

Neptune itself, rather than having been captured.[15] Perhaps the capture of Triton perturbed the orbit of this possibly pre-existing satellite into its highly elliptical path, which it still follows today?

Neptune's Inner Moons

Voyager 2's flypast would be an opportunity to discover more moons and even though Nereid's position and resulting images were disappointing, all was not lost, as the cameras would image a further six satellites, which would increase the known number of moons around Neptune threefold. The new moons would be called Naiad, Thalassa, Despina, Galatea and Proteus. As a group, their names were approved by the International Astronomical Union on 16 September 1989. Since Neptune was named after the Roman god of the sea, its moons are named after various lesser sea gods and nymphs from Greek mythology.

Proteus was the first new moon to be imaged, as it emerged from the glare of the planet, where it had been hidden, on 16 June 1989. Perhaps surprisingly, Proteus is the second-largest satellite of Neptune, with a diameter of 420 km. Its dark surface, which reflects less than 10 per cent of the sunlight it receives, and its orbital

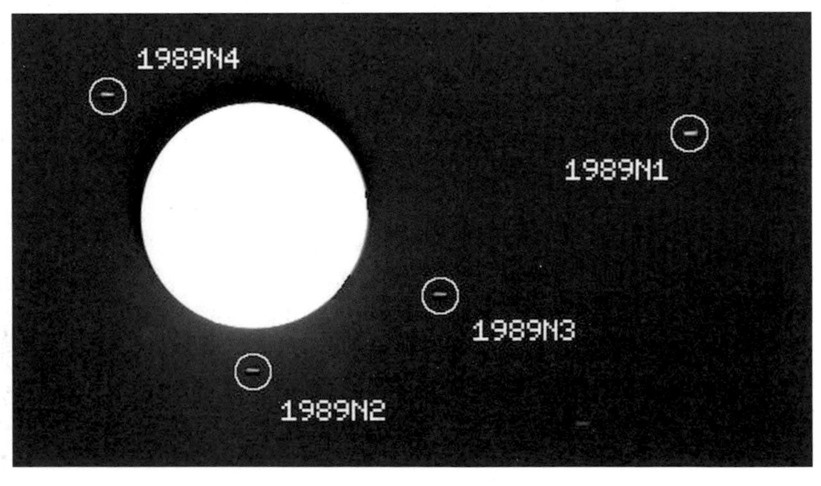

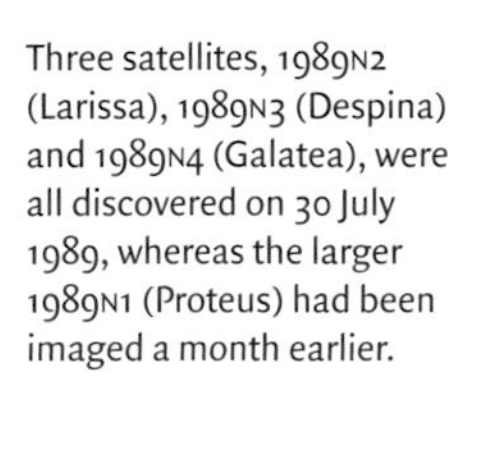
Three satellites, 1989N2 (Larissa), 1989N3 (Despina) and 1989N4 (Galatea), were all discovered on 30 July 1989, whereas the larger 1989N1 (Proteus) had been imaged a month earlier.

position, a relatively close 117,646 km from the surface of Neptune, meant it was hard to observe from Earth. Voyager imaged a moon that was asymmetrical, box-like in shape and heavily cratered. It is a second-generation moon, with material accreted together after the capture of Triton. Its diameter is about as large as a body can be without being pulled into a true spherical shape by the force of gravity. Due to the proximity of Voyager, which flew past at 146,000 km, the images reveal some interesting surface features. These include valleys, grooves and scarps. There is a large crater called Pharos; its diameter of 260 km is large when compared to the total size of the moon, namely 420 km. The heavily cratered surface shows that the surface is as old as the moon itself.

Proteus, 1989N1, half-illuminated by the Sun.

A similarly heavily cratered moon was first imaged around July 1989. Designated S/1989 N2, the team soon realized that this was the same moon found by Reitsema's team at the University of Arizona (S/1981 N1) in 1981 by stellar occultation observations. Larissa is likely to be a remnant or fragment from a larger body that was disrupted by Triton's capture. It is thought that Proteus and Larissa were once held in an orbital resonance of 1:2, where Larissa would make two orbits for every one of Proteus's, but this resonance no longer holds true. Proteus migrated outwards to its current location several hundred thousand years ago, most likely as a result of the pull of Triton on the system.

Neptune's moon Larissa imaged by Voyager 2, 24 August 1989.

The other four moons discovered by Voyager 2, in order of distance from the planet, are Naiad, Thalassa, Despina and

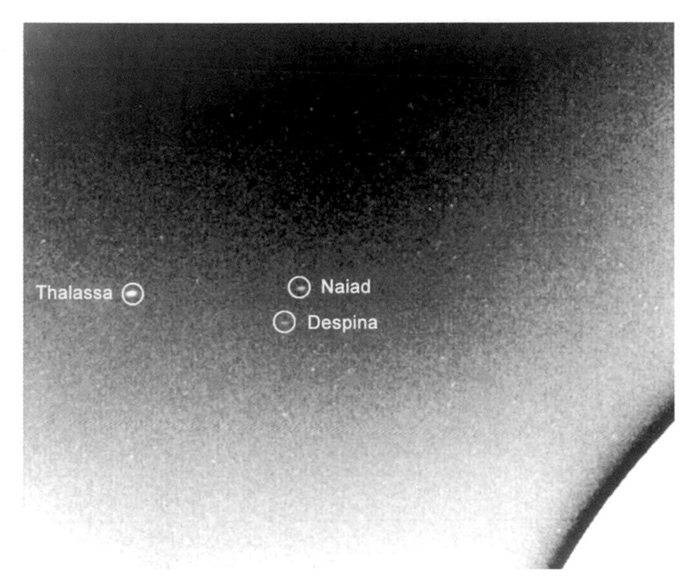

Despina (1989N3), Thalassa (1989N5) and Naiad (1989N6) are all spiralling towards the planet as their orbits decay due to tidal deceleration. The future of these satellites is limited. Imaged at a range of 5.9 million km.

Galatea. All these inner moons have a semi-major axis of 62,000 km or less from Neptune and have orbital periods of less than 0.5 days. Apart from Naiad they have almost no or little inclination to Neptune's orbital plane and they do not have great eccentricity. Due to their irregular shapes, they are all thought to have formed from accrued material and rubble amassed as a result of the capture of Triton. Naiad, the smallest, has an irregular shape of 96 × 60 × 52 km and is found closest to the planet. The largest, Galatea, with an average diameter of 204 × 184 × 144 km orbits furthest away from Neptune.

Between July 2002 and October 2003 the 10-metre Keck Telescope on Mauna Kea, Hawaii, was used to observe the inner moons surrounding Neptune. By using adaptive optics, all the moons, except for Naiad, were observed, allowing for new positional limitations to be set. Naiad was, in effect, lost; this small inner moon had disappeared. It would be observed again in images taken by the Hubble Space Telescope in 2004, which were re-examined in 2013. Astronomers were able to track its movement using eight archival images; the moon was found at a tiny 80 km distant to where it had been expected.

Naiad, Thalassa, Despina and Galatea all have an orbital period that is less than the rotational period of the planet. This means they are all gradually spiralling inwards due to tidal deceleration. They may eventually impact into the planet itself, spectacularly crashing into the upper atmosphere. Alternatively they may break up once

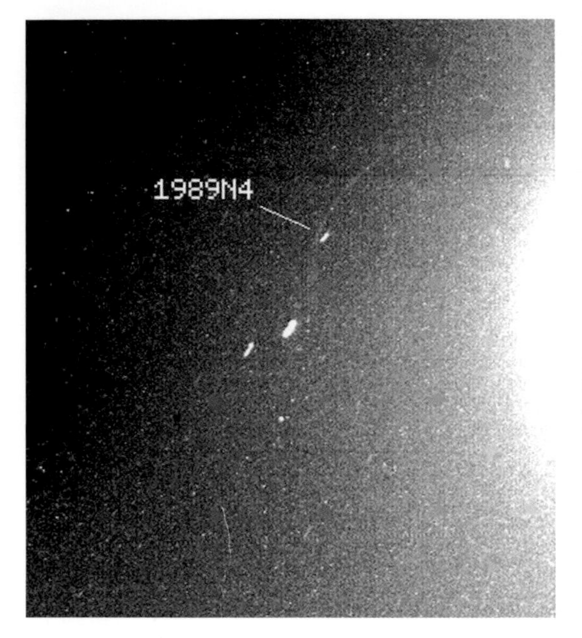

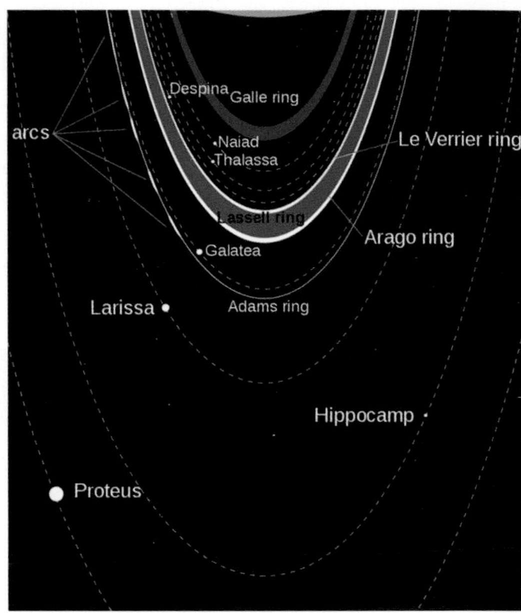

Neptune's moon Galatea (1989 N4) and new ring arcs, or partial rings.

Location of rings and arcs in comparison to inner moons.

they pass their Roche limit (this is the minimum distance a satellite can approach a planet without being pulled apart by the planet's gravity) and even go on to form a new planetary ring. This will not happen in unison and once Thalassa is pulled to pieces it will be only a matter of time before the debris affects the decreasing orbit of Despina. The deteriorating orbit of Despina will eventually have a crucial impact on the Le Verrier ring, disturbing it and adding material as it is pulled apart.

Neptune's Outer Moons

The next batch of discoveries of Neptunian moons were a result of ground-based telescope observations. These would all be outer moons, missed by Voyager because they were so far away from the spacecraft as to be too faint to see. In 2001, as a result of multiple discoveries of small bodies around the other gas giants, a team led by Mathew J. Holman turned the 4-metre telescope at the

The satellites' orbital characteristics: the vertical axis is the inclination, the horizontal the semi-major axis (the yellow line indicates eccentricity).

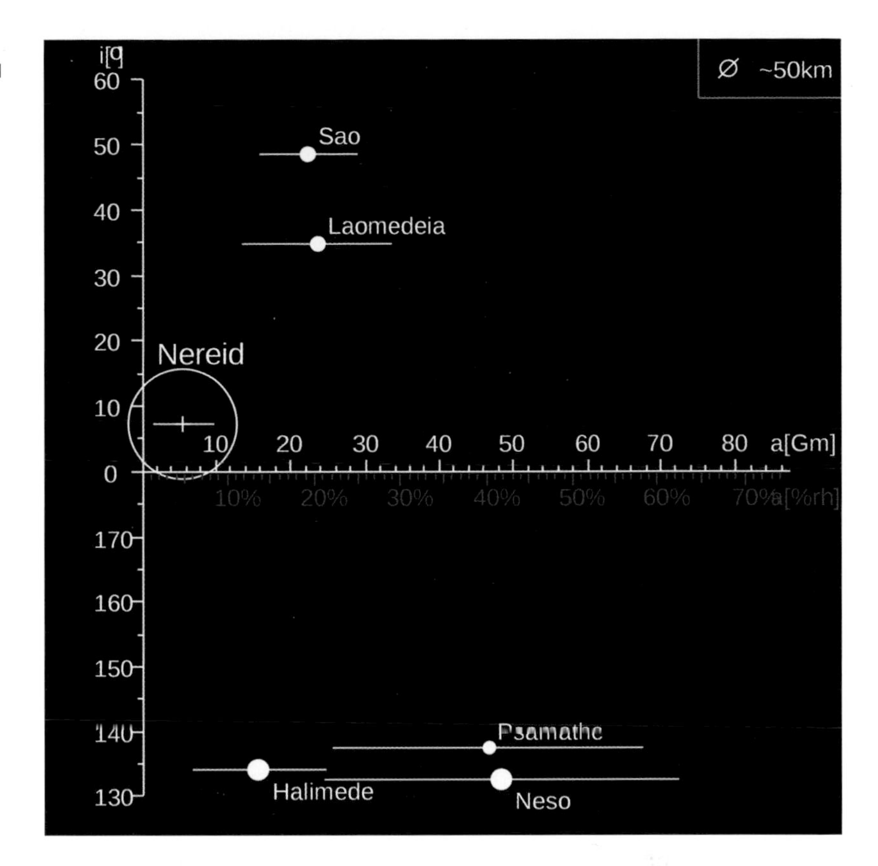

Cerro Tololo Inter-American Observatory in Chile and the 3.6-metre Canada-France-Hawaii Telescope in Hawaii to hunt for new candidates. Faint blips on images had to be tracked over a number of days before any positive identifications could be made. To substantiate any findings from this search, additional observations of the candidates had to be made using the 5-metre Hale Telescope on Mount Palomar and the 8-metre Very Large Telescope in Chile. In order of distance from the planet came Halimede, Sao, Laomedeia and Neso. They were quickly followed in the autumn of 2003 by the observation of Psamathe by a team led by Scott S. Sheppard using the Subaru Telescope located on Mauna Kea in Hawaii. All these moons orbit outside the path of Triton, at inclinations above and below 40 per cent of the orbital plane of Neptune. They are all considered to be captured objects resulting from Neptune's orbit close to the Kuiper Belt and are therefore classified as irregular moons. Their discovery is all the more gratifying, as they are very small, ranging from just 40 km to 62 km in diameter.

Sao and Laomedeia orbit in the same direction as Neptune, but the other three are retrograde. The similarity in orbits and the size of the eccentricity of Neso and Psamathe suggest they have a common origin and possibly come from a family of satellites created during the break-up of a larger moon. Neso orbits Neptune at the greatest distance of any moon discovered in the solar system, at an average distance of 48 million km extending to 72 million km at apocentre (the furthest point from the planet in its orbit); this is almost half the distance from the Sun to the Earth.

Hippocamp

The most recent moon to be discovered was spotted by Mark Robert Showalter and a team at the SETI Institute, California. Looking back through Hubble Space Telescope images dating back to 2009, they discovered a small white dot which looked like a moon. Turning to the archives, they were able to use 150 archive images taken from the Hubble Space Telescope that showed the object as early as 2004 and this allowed them to plot its orbital characteristics.

The moon itself is just 34 km in diameter and located between the orbits of Proteus and Larissa. It is located remarkably close to Proteus, at a distance of only 12,000 km from its neighbouring satellite. It is strange to have such a small object next to a much larger inner moon – usually a smaller moon would be ejected from its orbit by the larger one, or it might crash into the much larger object. This has led scientists to conclude that Hippocamp may be a third-generation moon, or a chip off the much-larger Proteus's block. The cataclysmic event, which tore through the Neptune

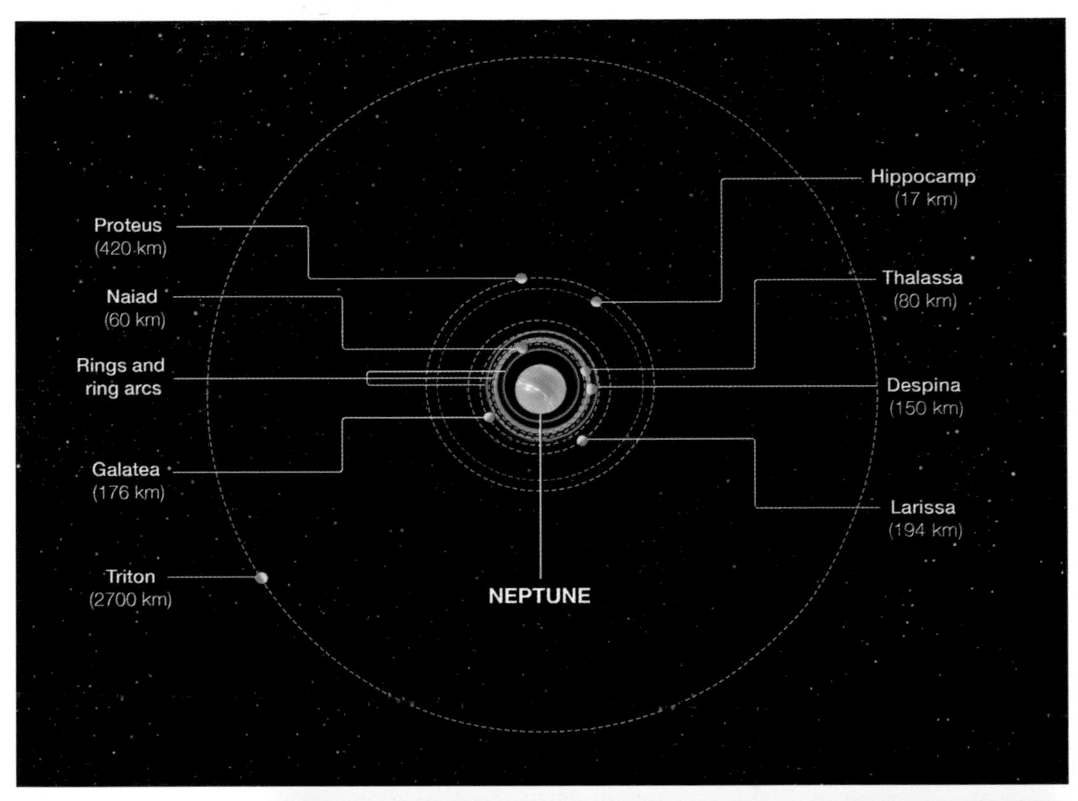

Artist's impression of Neptune and its inner moons, 19 February 2019.

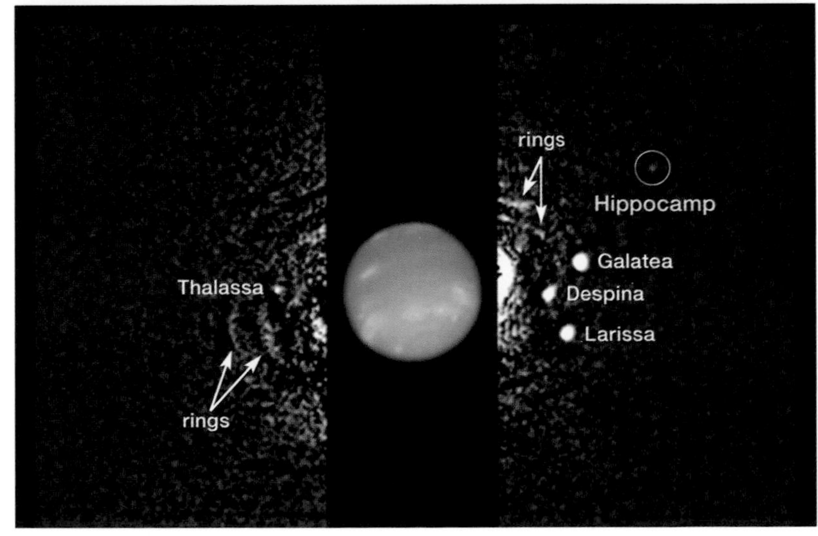

Hippocamp taken by the Hubble Space Telescope's Wide Field Camera 3 in visible light, 19 August 2009.

Neptune and an artist's depiction of its newest family member, Hippocamp, 19 February 2019.

system when Triton was captured and led to the creation of Proteus and other moons, was only the beginning in creating this dynamic system. This was to be followed by a heavy bombardment of comets, and one of these is suggested to have hit Proteus, splitting off the smaller fragment which is now called Hippocamp.[16] The large crater Pharos, seen on the surface of Proteus, may be evidence of where Hippocamp originated from. This could be the region where the moon broke off. It would have taken a large impact to achieve this fracture, which is why Showalter and his team suggest that a comet was the most likely cause. A return mission to the Neptunian system will probably bring further moons to light.

Future Missions

Trident

The Trident mission was proposed as part of NASA's Discovery Program 2019. This mission was to focus on Triton, Neptune's largest moon. The mission was selected for further investigation, along with three further missions, but unfortunately it did not get chosen. The JPL Trident project scientist Karl Mitchell explains why it is critical to return to the moon: 'As we said to NASA in our mission proposal, Triton isn't just a key to solar system science – it's a whole keyring: a captured Kuiper Belt object that evolved, a potential ocean world with active plumes, an energetic ionosphere and a young, unique surface.'[17]

If Trident had got the go-ahead, it would have launched in 2025 or 2026 using a once-in-a-thirteen-year window where Earth is aligned with Jupiter. It would allow a gravity assist from Jupiter, making the craft reach Triton in 2038. On arrival it would have made observations for thirteen days. It had been planned to map the surface in full, allowing for observations to be made of the icy shell and investigating evidence of the subsurface ocean which constantly renews the surface. Voyager 2 spotted plumes which

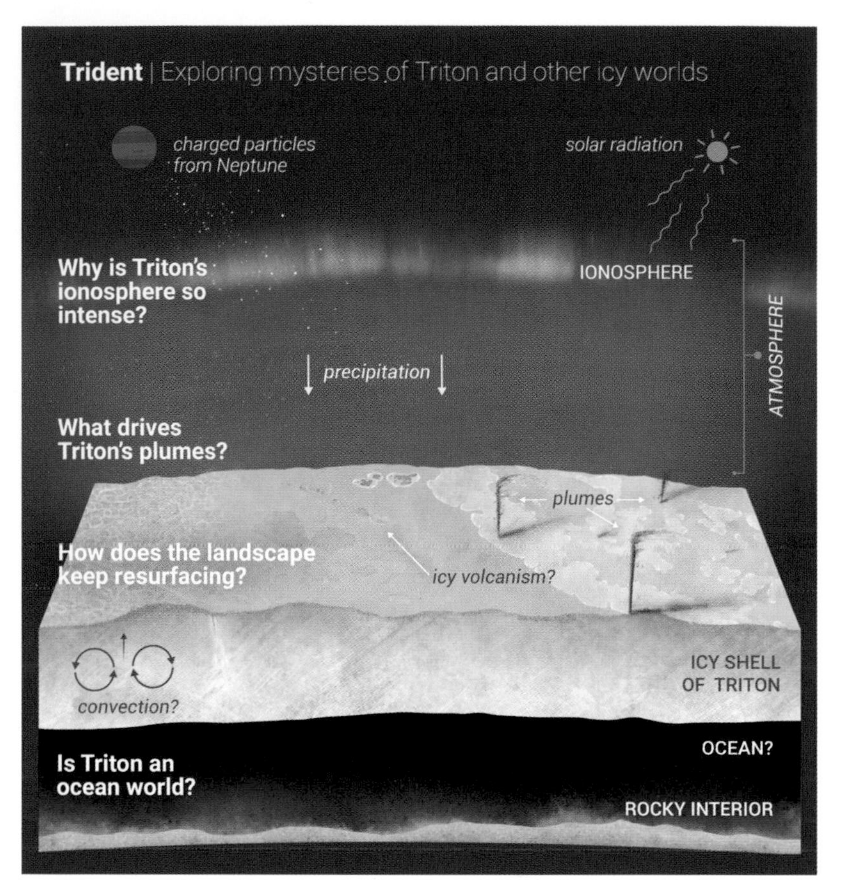

The NASA mission Trident would have explored Triton, the large moon of Neptune, by examining its habitability, mapping its surface and investigating how it keeps renewing itself.

refresh the surface, making it one of the youngest in the solar system, and it is believed that these could originate from an ocean under the crust. The mission would have examined the plumes and tried to draw conclusions about the processes driving the activity – was it icy vulcanism originating from a subsurface ocean? Or could it be the result of a greenhouse process? This part of the mission was time-sensitive and if the craft was to reach the moon any later than 2040 the Sun would have moved too far to the north and the team would not be able to test what was powering the plumes. It will take another hundred years for the opportunity to present itself again.[18]

The craft would have flown through the thin atmosphere, reaching heights of just 500 km above the surface, and taken samples from the ionosphere. The ionosphere is far more intense than others measured in the solar system, which is a mystery, as ionospheres are driven by the Sun, and Neptune and Triton are much further away from its influence than bodies with less intense ionospheres. The mission would have offered real insight into the habitability of icy moons and the characteristics that make similar bodies candidates for life in the solar system.

Triton Hopper

The Triton Hopper is a concept for a lander which is currently being developed by NASA. It was proposed in 2015 and by March 2018

The Triton Hopper, a concept lander from NASA.

Artist's impression of Neptune from the surface of Triton, the largest moon to orbit the planet.

it was in Phase II at the NASA Institute for Advanced Concepts. It is being designed to land on the surface of Triton; as a 'hopper' it could land and fly over a number of terrains, possibly even collecting material from the plumes as it flew through them. It would utilize the moon's nitrogen ice as a propellant and could make 'hops' up to 300 km in length, allowing for long-distance exploration of the surface.[19] The advantage of a lander is that it could collect samples by drilling into the surface and then conduct chemical analysis of these samples while photographing it all in close proximity. The development of this lander would allow for *in situ* exploration on what is considered one of the most fascinating objects within the solar system.

NINE
Observing the Ice Giants

You may wish to view the ice giants in the night sky yourself. Before heading out, it is worth considering what to expect when observing these planets, as they can be difficult targets. Uranus, at double the distance from Earth than Saturn, is the less challenging of the two, being at the limits of naked-eye observation in a dark sky area, and armed with a suitable finder chart and binoculars, it should be relatively easy to find in comparison to Neptune, which is dim, distant and cannot be seen without optical equipment.

Uranus

With a peak magnitude of around 5.7 it is on the outer limits of what we are capable of observing with the naked eye, even in an area free from light pollution. There are tips that can help you spot this greenish-blue planet in the sky. It is best to observe the planet at opposition, although it can be observed either side of opposition. At opposition as the Sun sets in the west, the planet will be rising in the east; the planet will reach its highest altitude at local midnight. As Uranus is directly opposite the Sun in the sky, it will appear at its highest altitude and its brightest, having an apparent magnitude of 5.64. You will need to be in a dark location, away from any ambient light, and choose a clear, moonless night to observe Uranus with the naked eye. Without optical aid it will just appear as a point of light,

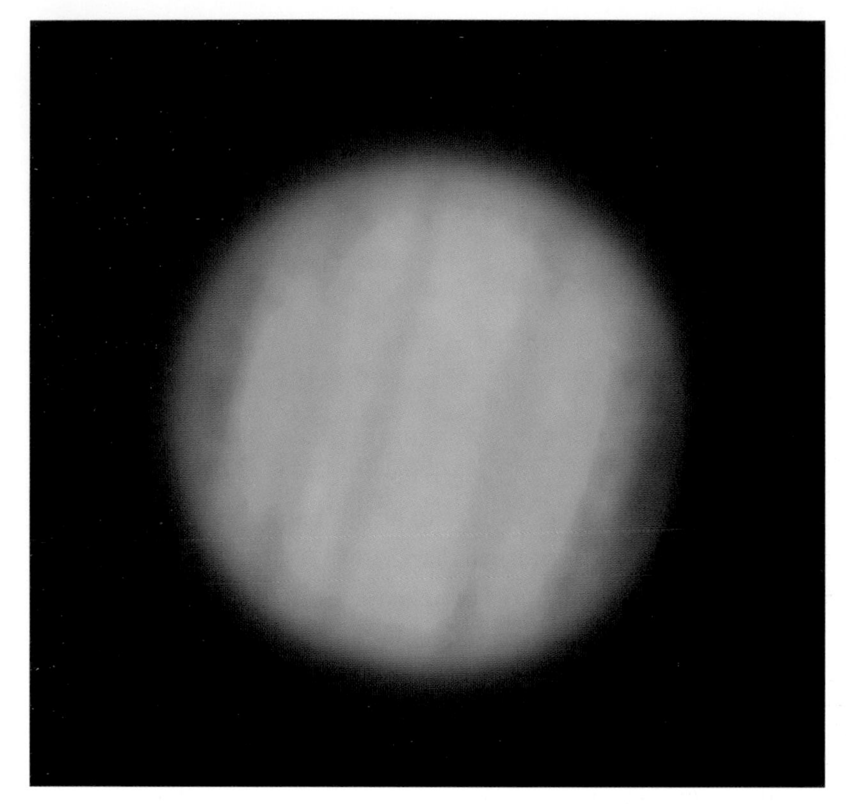

Uranus, 27 November 2018, 1-m telescope. ASI174MM camera.

Banding was clearly visible on Uranus in this colour drawing made by Martin Stagel on 11 November 2008, using a C14 telescope with a magnification of 260x. The drawing was made between 19:36 and 19:50 UT under excellent seeing conditions, with the planet 36° above the horizon. No filters were used, in order to draw the planet in real colours.

like the surrounding stars, so it is essential to find out which point of light is the planet, by using a star chart or planetarium software.

When binoculars or telescopes are turned to Uranus it looks like a greenish-bluish object. Turning a 70-millimetre telescope on the planet will enable you to barely resolve the disc. Larger telescopes of 320 mm and above will allow you to see a better view of the disc and the five largest moons orbiting the planet. Many amateurs now use equipment with CCD imagers or DSLR cameras which enable them to image the planet and thus see more detail. These images can draw out the detail of atmospheric bands, particularly the northern region, which is turning towards us as the planet approaches its winter solstice in 2028. Observations of this kind help to increase our knowledge about the weather patterns at this planet.

The rings of Uranus are often described as tenuous; can amateur astronomers observe the ring system with the modern technology that is available to amateurs? Amateur astronomers have reported being able to image the rings around the planet with modern CCD imaging set-ups. The ring system was imaged by the leading British astrophotographer Damien Peach in 2015. The use of different wavelengths to image the planet can help bring out features. Using an infrared filter can bring out even more detail on the surface and is used alongside visible light to show the planet in all its glory.

There is even the potential to image the planet in broad daylight. Austrian planetary observer Martin Stangl explains:

> I wondered if it would be possible to photograph Uranus when the sun is up, so I followed the planet until after sunrise. It was very low contrast of course, but I managed to have the planet on the screen until about 20 minutes after sunrise. The sun was shining brightly on my setup when I could still see Uranus on the screen. An absolutely incredible sight. But the biggest surprise came when I processed the images. In the final composite, after working with several free and commercial software [sic], I managed to even bring out atmospheric details on the small disc. As they are the same as earlier images during the night hours, there is no doubt they are real and not artefacts.[1]

A daylight photograph from 16 August 2018. It was stacked from 3,500 single video frames made with a ZWO ASI 178MM monochrome camera at f/27 through the red RGB filter between 09:51 and 10:10 UT, which corresponds to a mean of 18 minutes after local sunrise. Every single frame used for the stack was after sunrise, to produce a 100 per cent daylight shot.

Neptune

It took well over two hundred years from the first use of a telescope to recognize this point of light as a planet. Its average distance is 30 AU from the Sun and even at opposition in 2021 it was still 28.92 AU from the Sun. Neptune currently has an apparent magnitude that ranges from 7.67 to 7.89. As it reaches opposition it will brighten in magnitude. Although the magnitude of the planet does not vary by much over the year, the brightness has increased by 10 per cent since the 1980s, tapering off around 2000.[2] The apparent magnitude of Neptune means you will need a telescope or binoculars to observe it. The disc is between 2.2 and 2.4 arcseconds in size and it appears as a blue spot of light in smaller telescopes. Amateur astronomers using binoculars or small telescopes may find it helpful to use a pointer star when looking for the planet, otherwise they can find identifying the planet from the background of stars a tricky task; using planetarium software or a star chart will help you to identify the planet among the background stars. Observers using a 150-millimetre telescope will be able to see Neptune's disc. If larger telescopes (of over 300 mm in aperture) are used, details may be revealed on the disc. Observers who image this planet often try to get the brightest of its moons, Triton, into the shot. The use of filters, particularly in the infrared, allows for more detail and features to appear in the images. Taking photographs of this dim object requires patience. Imaging it requires much longer exposure times than the other planets, but as equipment capabilities improve this exposure time gets reduced. Mike Foulkes of the British Astronomical Association explains:

> Professional and amateur observation of the planets has benefited from digital imaging as have the observation of other astronomical objects. Producing good planetary images does depend upon a number of factors including: the telescope size, the seeing, the image scale, the exposure of individual frames, the number of frames captured, the camera, the filters used and the subsequent image processing. Obtaining a good focus is also important. Imaging the disks of Uranus and Neptune do [*sic*] benefit from the use of larger aperture telescopes due to their lower surface brightness and small apparent angular diameters compared to the other planets.[3]

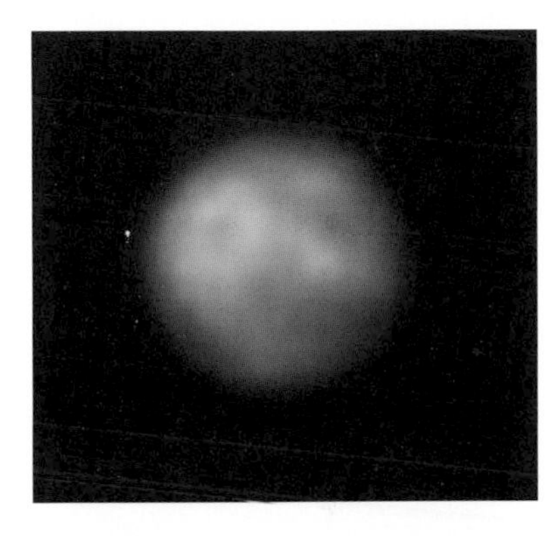

Martin Stangl, alongside Rolf Winkler, imaged Neptune on 15 August 2018 using a Celestron C14 Schmidt-Cassegrain telescope. The black-and-white image shows how detail can be teased out, with a lighter band extending round the equatorial region.

Damien Peach imaged a similar marking on Neptune in 2015. Shining brightly against the blue colour of the planet, the white streak can be seen extending across the equatorial belt region. The planet's largest moon, Triton, is also visible to the top right of the planet. He achieved this using a Celestron C14 telescope with filters and a CCD camera attached. He is, however, one of the finest amateur astrophotographers in the world, taking images that are as good as ones taken by large professional telescopes and so showing how the skill in taking images can overcome the limitations of a telescope's aperture.

As they are monitoring these two planets continually, the amateur community can stand alongside their professional colleagues at the forefront of making discoveries about Uranus and Neptune. They are now in a position to add observations of real value to our understanding of these distant ice giants.

Neptune, 15 August 2018. Stacked from 14,000 single video frames made with a ZWO ASI 178MM monochrome camera at f/11 through a Schott IR685 near-infrared filter between 05:40 and 06:32 UT.

Neptune, imaged on 19–20 September 2015, taken using a C14 telescope with an ASI224 camera and RG610 and IR685 filters.

Appendix I:

Future Time and Dates and Apparent Magnitude of the Planets at Opposition

Uranus

9 Nov 2022 AT 08:12 GMT (08:12 UTC), 5.64 Apparent Magnitude
13 Nov 2023 AT 17:06 GMT (17:06 UTC), 5.63 Apparent Magnitude
17 Nov 2024 AT 02:30 GMT (02:30 UTC), 5.61 Apparent Magnitude
21 Nov 2025 AT 12:11 GMT (12:11 UTC), 5.60 Apparent Magnitude
25 Nov 2026 AT 22:27 GMT (22:27 UTC), 5.58 Apparent Magnitude
30 Nov 2027 AT 09:07 GMT (09:07 UTC), 5.57 Apparent Magnitude
3 Dec 2028 AT 20:14 GMT (20:14 UTC), 5.55 Apparent Magnitude
8 Dec 2029 AT 07:57 GMT (07:57 UTC), 5.53 Apparent Magnitude
12 Dec 2030 AT 20:21 GMT (20:21 UTC), 5.52 Apparent Magnitude

Neptune

16 Sep 2022 AT 23:05 BST (22:05 UTC), 7.81 Apparent Magnitude
19 Sep 2023 AT 12:02 BST (11:02 UTC), 7.81 Apparent Magnitude
21 Sep 2024 AT 01:01 BST (00:01 UTC), 7.81 Apparent Magnitude
23 Sep 2025 AT 13:38 BST (12:38 UTC), 7.81 Apparent Magnitude
26 Sep 2026 AT 02:21 BST (01:21 UTC), 7.81 Apparent Magnitude
28 Sep 2027 AT 15:03 BST (14:03 UTC), 7.81 Apparent Magnitude
30 Sep 2028 AT 03:33 BST (02:33 UTC), 7.81 Apparent Magnitude
2 Oct 2029 AT 16:09 BST (15:09 UTC), 7.81 Apparent Magnitude
5 Oct 2030 AT 04:30 BST (03:30 UTC), 7.81 Apparent Magnitude

APPENDIX II:

URANUS DATA

Orbital Characteristics

epoch	J2000
aphelion	3,008 million km (20.11 AU)
perihelion	2,742 million km (18.33 AU)
semi-major axis	2,875.04 million km (19.2184 AU)
eccentricity	0.046381
orbital period	
84.0205 years	
30,688.5 days	
42,718 Uranian solar days	
synodic period	369.66 days
average orbital speed	6.80 km/sec
mean anomaly	142.238600°
inclination	
0.773° to ecliptic	
6.48° to Sun's equator	
1.02° to invariable plane	
longitude of ascending node	74.006°
time of perihelion	19 August 2050
argument of perihelion	96.998857°

Physical Characteristics

mean radius 25,362±7 km
equatorial radius 25,559±4 km (4.007 Earths)
polar radius 24,973±20 km (3.929 Earths)
flattening 0.0229±0.0008
circumference 159,354.1 km
surface area 8.1156×109 km^2 (15.91 Earths)
volume 6.833×10^{13} km^3 (63.086 Earths)
mass $(8.6810 \pm 0.0013) \times 10^{25}$ kg (14.536 Earths)
GM 5,793,939±13 km^3/s^2
mean density 1.27 g/cm^3
surface gravity 8.69 m/s^2
  0.886 g
moment of inertia factor 0.23 I/MR^2 (estimate)
escape velocity 21.3 km/sec
sidereal rotation period −0.71833 days (retrograde)
  17 hr 14 min 24
equatorial rotation velocity 2.59 km/sec (9,320 km/hr)
axial tilt 97.77° (to orbit)
North pole right ascension 257.311° : 17 hr 9 min 15 sec
North pole declination −15.175°
albedo
  0.300 (Bond)
  0.488 (geometric)
surface temp. (min mean max)
  1 bar 76 K (−197.2 °C)
  0.1 bar 47 K, 53 K, 57 K (tropopause)
apparent magnitude 5.38 to 6.03
angular diameter 3.3 ″ to 4.1 ″
atmosphere
  scale height 27.7 km

Composition by Volume (Below 1.3 bar)

83 ± 3 per cent hydrogen (H_2)
15 ± 3 per cent helium (He)
2.3 per cent methane (CH_4)
0.009 per cent (0.007–0.015 per cent) hydrogen deuteride (HD)
hydrogen sulphide (H_2S)

Ices
- ammonia (NH_3)
- water (H_2O)
- ammonium hydrosulphide (NH_4SH)
- methane hydrate

Appendix III:

Uranus Ring Data

Rings of Uranus	Radius (km)	Radius/ Eq. radius	Optical Depth	Albedo (x 10^{-3})	Width (km)	Eccentricity
Uranus Equator	25,559	1				
6	41,837	1.637	~0.3	~15	1.5	0.001
5	42,234	1.652	~0.5	~15	~2	0.0019
4	42,571	1.666	~0.3	~15	~2	0.0011
Alpha	44,718	1.75	~0.4	~15	04-Oct	0.0008
Beta	45,661	1.786	~0.3	~15	05-Nov	0.0004
Eta	47,176	1.834	~0.4-	~15	1.6	0.001
Gamma	47,627	1.863	~0.3+	~15	01-Apr	0.0011
Delta	48,300	1.9	~0.5	~15	03-Jul	0.0004
Lambda	50,024	1.957	~0.1	~15	~2	0
Epsilon	51,149	2.006	0.5–2.3	~18	20–96	0.0079
R/2003 U2	67,300	2.633	<<0.01		3,800	0
R/2003 U1	97,700	3.823	<<0.01		17,000	0

Credit: NASA. Dr David R. Williams

Appendix IV: Neptune Data

Orbital Characteristics

epoch	J2000
aphelion	4.54 billion km (30.33 AU)
perihelion	4.46 billion km (29.81 AU)
semi-major axis	4.50 billion km (30.07 AU)
eccentricity	0.008678
orbital period	
164.8 years	
60,182 days	
89,666 Neptunian solar days	
synodic period	367.49 days
average orbital speed	5.43 km/sec
mean anomaly	256.228°
inclination	
1.767975° to ecliptic	
6.43° to Sun's equator	
0.72° to invariable plane	
longitude of ascending node	131.784°
time of perihelion	4 September 2041
argument of perihelion	276.336°

Physical Characteristics

mean radius	24,622±19 km
equatorial radius	24,764±15 km (3.883 Earths)
polar radius	24,341±30 km (3.829 Earths)
flattening	0.0171±0.0013
surface area	7.6183×109 km^2 (14.98 Earths)
volume	6.254×1013 km^3 (57.74 Earths)
mass	1.02413×10^{26} kg
	17.147 Earths
	5.15×10^{-5} Suns
mean	
1.638 g/cm^3	
surface gravity	
11.15 m/s^2	
1.14 g	
moment of inertia factor	0.23 (estimate)
escape velocity	23.5 km/sec
sidereal rotation period	0.6713 days 16 hr 6 min 36 sec
equatorial rotation velocity	2.68 km/sec (9,650 km/hr)
axial tilt	28.32° (to orbit)
North pole right ascension	19 hr 57 min 20 sec 299.3°
North pole declination	42.950°
albedo	
0.290 (bond)	
0.442 (geom.)	
surface temp. (min mean max)	
1 bar level	72 K (−201°C)
0.1 bar (10 kPa)	55 K (−218°C)
apparent magnitude	7.67 to 8.00
angular diameter	2.2″–2.4″
atmosphere	
scale height 19.7±0.6 km	

Composition by Volume

Gases

80 per cent±3.2 per cent hydrogen (H_2)
19 per cent±3.2 per cent helium (He)
1.5 per cent±0.5 per cent methane (CH_4)
~0.019 per cent hydrogen deuteride (HD)
~0.00015 per cent ethane (C_2H_6)

Ices

ammonia (NH_3)
water (H_2O)
ammonium hydrosulphide (NH_4SH)
methane ice (?) ($CH_4 \cdot 5.75H_2O$)

Appendix V:

Neptune Ring Data

	Radius (km)	Radius Eq. radius	Optical Depth	Albedo (x 10^{-3})	Width (km)
Neptune Equator	24,766	1			
Galle (1989N3R)	~41,900	1.692	~0.0001	~15	~2,000
Le Verrier (1989N2R)	~53,200	2.148	~0.01	~15	< 100
Lassell (1989N4R)	~53,200	2.148	~0.0001	~15	~4,000
Arago (1989N4R)	~57,200	2.31	< 100		
Unnamed (indistinct)	61,953	2.501			
Adams (1989N1R)	62,933	2.541	0.01–0.1	~15	~15
Arcs in Adams Ring:					
Courage	62,933	2.541	0.12		~15
Liberté	62,933	2.541	0.12		~15
Égalité 1	62,933	2.541	0.12	~40	~15
Égalité 2	62,933	2.541	0.12	~40	~15
Fraternité	62,933	2.541	0.12		~15

Credit: NASA. Dr David R. Williams

APPENDIX VI: SATELLITES OF URANUS

Name	Diameter (km)	Semi-Major Axis (km)	Orbital Period (Days)	Inclination (°)	Eccentricity	Discovery Year	Discoverer	
VI Cordelia	40 ± 6 (50 x 36)	49,770	0.33503	0.08479°	0.00026	1986	Terrile	(Voyager 2)
VII Ophelia	43 ± 8 (54 x 38)	53,790	0.3764	0.1036°	0.00992	1986	Terrile	(Voyager 2)
VIII Bianca	51 ± 4 (64 x 46)	59,170	0.43458	0.193°	0.00092	1986	Smith	(Voyager 2)
IX Cressida	80 ± 4 (92 x 74)	61,780	0.46357	0.006°	0.00036	1986	Synnott	(Voyager 2)
X Desdemona	64 ± 8 (90 x 54)	62,680	0.47365	0.11125°	0.00013	1986	Synnott	(Voyager 2)
XI Juliet	94 ± 8 (150 x 74)	64,350	0.49307	0.065°	0.00066	1986	Synnott	(Voyager 2)
XII Portia	135 ± 8 (156 x 126)	66,090	0.5132	0.059°	0.00005	1986	Synnott	(Voyager 2)
XIII Rosalind	72 ± 12	69,940	0.55846	0.279°	0.00011	1986	Synnott	(Voyager 2)
XXVII Cupid	≈ 18	74,800	0.618	0.100°	0.0013	2003	Showalter and Lissauer	
XIV Belinda	90 ± 16 (128 x 64)	75,260	0.62353	0.031°	0.00007	1986	Synnott	(Voyager 2)
XXV Perdita	30 ± 6	76,400	0.638	0.0°	0.0012	1999	Karkoschka	(Voyager 2)
XV Puck	162 ± 4	86,010	0.76183	0.3192°	0.00012	1985	Synnott	(Voyager 2)
XXVI Mab	≈ 25	97,700	0.923	0.1335°	0.0025	2003	Showalter and Lissauer	
V Miranda	471.6 ± 1.4 (481 x 468 x 466)	129,390	1.41348	4.232°	0.0013	1948	Kuiper	

I Ariel	1157.8±1.2 (1162 x 1156 x 1155)	191020	2.52038	0.260°	0.0012	1851	Lassell	
II Umbriel	1169.4±5.6	266300	4.14418	0.205°	0.0039	1851	Lassell	
III Titania	1576.8±1.2	435910	8.70587	0.340°	0.0011	1787	Herschel	
IV Oberon	1522.8±5.2	583520	13.4632	0.058°	0.0014	1787	Herschel	
XXII Francisco	≈ 22	4282900	−267.09	147.250°	0.1324	2003	Holman et al.	
XVI Caliban	42+20	7231100	−579.73	141.529°	0.1812	1997	Gladman et al.	
XX Stephano	≈ 32	8007400	−677.47	143.819°	0.2248	1999	Gladman et al.	
XXI Trinculo	≈ 18	8505200	−749.40	166.971°	0.2194	2001	Holman et al.	
XVII Sycorax	157+23	12179400	−1288.38	159.420°	0.5219	1997	Nicholson et al.	
XXIII Margaret	≈ 20	14146700	1661	57.367°	0.6772	2003	Sheppard and Jewitt	
XVIII Prospero	≈ 50	16276800	−1978.37	151.830°	0.4445	1999	Holman et al.	
XIX Setebos	≈ 48	17420400	−2225.08	158.235°	0.5908	1999	Kavelaars et al.	
XXIV Ferdinand	≈ 20	20430000	−2790.03	169.793°	0.3993	2003	Holman et al.	

Appendix VII:

Satellites of Neptune

Name	Diameter (km)	Semi-Major Axis (km)	Orbital period (Days)	Orbital inclination (°)	Eccentricity	Discovery Year	Discoverer
III Naiad	60.4 (96 × 60 × 52)	48,224	0.2944	4.691	0.0047	1989	Voyager Science Team
IV Thalassa	81.4 (108 × 100 × 52)	50,074	0.3115	0.135	0.0018	1989	Voyager Science Team
V Despina	156 (180 × 148 × 128)	52,526	0.3346	0.068	0.0004	1989	Voyager Science Team
VI Galatea	174.8 (204 × 184 × 144)	61,953	0.4287	0.034	0.0001	1989	Voyager Science Team
VII Larissa	194 (216 × 204 × 168)	73,548	0.5555	0.205	0.0012	1981	Reitsema et a
XIV Hippocamp	34.8±4.0	105,283	0.95	0.064	0.0005	2013	Showalter et al.
VIII Proteus	420 (436 × 416 × 402)	117,646	1.1223	0.075	0.0005	1989	Voyager Science Team
I Triton	2705.2±4.8	354,759	−5.8769	156.865	0	1846	Lassell
II Nereid	357 ± 13	5,513,800	360.13	7.09	0.7507	1949	Kuiper
IX Halimede	≈ 62	16,681,000	−1879.33	112.898	0.2909	2002	Holman et al
XI Sao	≈ 44	22,619,000	2919.16	49.907	0.2827	2002	Holman et al
XII Laomedeia	≈ 42	23,613,000	3175.62	34.049	0.4339	2002	Holman et al
X Psamathe	≈ 40	46,705,000	−9128.74	137.679	0.4617	2003	Sheppard et a
XIII Neso	≈ 60	50,258,000	−9880.63	131.265	0.4243	2002	Holman et al

References

1 Uranus, Its Discovery and Pre-Voyager Observations

1 William Herschel, 'Account of a Comet', *Scientific Papers* (London, 1912), vol. i, p. 30.
2 Constance A. Lubbock, *The Herschel Chronicle* (Cambridge, 1933).
3 Herschel, *Scientific Papers*, vol. i, p. xxx.
4 Ibid., p. 95.
5 Owen Gingerich, 'The Naming of Uranus and Neptune', *Astronomical Society of the Pacific Leaflets*, viii/352 (1958), p. 9.
6 William Sheehan et al., *Neptune: From Grand Discovery to a World Revealed: Essays on the 200th Anniversary of the Birth of John Couch Adams* (London, 2021).
7 Ibid.
8 Ibid., p. 56.
9 Herschel, *Scientific Papers*, vol. i, p. 102.
10 Wes Lockwood, 'A Century of Uranus Observation at Lowell', *Lowell Observer* (Fall 2010).
11 William Herschel, 'Observations and Reports Tending to the Discovery of One or More Rings of the Georgian Planet, and the Flattening of Its Polar Regions', *Philosophical Transactions* (1798), pp. 67, 88.
12 Richard Baum, *The Planets: Some Myths and Realities* (Frome, 1973), pp. 114–19.
13 Stuart Eves, 'Did William Herschel Discover the Rings of Uranus in the 18th Century?', Royal Astronomical Society, www.spaceref.com, 24 April 2007.
14 Ibid.
15 Father Angelo Secchi, Letter to the French Académie des sciences dated 20 March 1869; Arthur Francis O'Donel Alexander, *The Planet Uranus: A History of Observation, Theory and Discovery* (New York, 1965), p. 151.
16 Thomas Hughes Buffham, 'Markings Observed on Uranus', *Monthly Notices of the Royal Astronomical Society*, xxxiii (1873), p. 164.

17 Alexander, *The Planet Uranus*, pp. 134–5.
18 Kevin Bailey, 'Thomas Hughes Buffam: Uranus Pioneer', *Journal of the British Astronomical Association*, CXXVII/4 (2017), pp. 207–8.
19 Giovanni Schiaparelli, 'Urano', *Astronomische Nachrichten*, CVI/2526 (1884), col. 86.
20 Richard M. Baum and Andrew J. Hollis, 'Uranus: The View from Earth', *Journal of the British Astronomical Association*, XCVI/2 (February 1986), p. 65.
21 Martin Mobberley, 'Reginald Lawson Waterfield, 1900–86, Eclipse Chaser and Comet Photographer Extraordinaire, Part I', *Journal of the British Astronomical Association*, CXXXI/3 (1921), pp. 158–70.
22 Michael James Hendrie, 'Reginald Lawson Waterfield, 1900–86', *Journal of the British Astronomical Association*, XCVII/4 (1987), pp. 211–14.
23 Patrick Moore, *New Guide to the Planets* (London, 1993), p. 163.
24 Vesto M. Slipher, 'Detection of the Rotation of Uranus', *Lowell Observatory Bulletin*, vol. II (1912), pp. 19–20.
25 Garry Hunt and Patrick Moore, *Atlas of Uranus* (Cambridge, 1989), p. 32.

2 Voyager 2 Flypast of Uranus

1 Alan Stern and David Grinspoon, *Chasing New Horizons: Inside the Epic First Mission to Pluto* (London, 2018), p. 12.
2 Eric Burgess, *Far Encounter: The Neptune System* (New York, 1991), pp. 1–2; 'Interview: Bradford Smith', *Space World* (November 1985), 5, record no. 005586, NASA Historical Reference Collection.
3 Carl Sagan, *The Cosmic Connection* (New York, 1973).
4 E. Myles Standish, 'Planet X: No Dynamical Evidence in the Optical Observations', *Astronomical Journal*, CV/5 (1993), pp. 2000–2006.
5 Andrew J. Butrica, 'Voyager: The Grand Tour of Big Science', in *From Engineering Science to Big Science: The NACA and NASA Collier Trophy Research Project Winners*, ed. Pamela E. Mack (Washington, DC, 1998), p. 270.
6 Garry Hunt and Patrick Moore, *Atlas of Uranus* (Cambridge, 1989), p. 52.
7 Ellis D. Miner, *Uranus: The Planet, Rings and Satellites* (London, 1990), pp. 270–72.
8 Ed Stone, 'Celebrating 30 Years of Voyager 2's Uranus Flyby', *R&D World* (25 January 2016).
9 Alan Cooper et al., 'A Global Environmental Crisis 42,000 Years Ago', *Science*, CCCVXXI/6531 (February 2021), pp. 811–18.
10 'Voyager 2 Discovers Clouds, Winds in Uranus Atmosphere', *Bangor Daily News* (23 January 1986), p. 10.
11 Andrew P. Ingersoll, 'Uranus', *Scientific American*, CCLVI/1 (January 1987), pp. 38–45.

12 Leonard Tyler et al., 'Voyager 2 Radio Science Observations of the Uranian System: Atmosphere, Rings, and Satellites', *Science*, CCXXXIII/4759 (July 1986), pp. 79–84.
13 Butrica, 'Voyager', p. 264.
14 Mark Littman, *Planets Beyond: Discovering the Outer Solar System* (New York, 1990).
15 Alexandra Witze, 'Gas Giant Spins a Surprise', *Nature* (12 November 2014), www.nature.com, accessed March 2021.
16 Gina A. DiBraccio and Daniel J. Gershman, 'Voyager 2 Constraints on Plasmoid-Based Transport at Uranus', *Geophysical Research Letters*, XLVI/19 (16 October 2019), pp. 10710–18.

3 Uranus after Voyager

1 'NASA's Hubble Discovers New Rings and Moons around Uranus' (22 December 2005), Release ID: 2005-33, www.hubblesite.org.
2 Imke de Pater, 'Faint New Ring Discovered around Uranus', *Keck Observatory News* (22 December 2005), www.keckobservatory.org, accessed 21 March 2021.
3 Edward Molter et al., 'Thermal Emission from the Uranian Ring System', *Astronomical Journal*, CLVIII/1 (July 2019), p. 8.
4 'Astronomers See "Warm" Glow of Uranus's Rings' (20 June 2019), www.eurekalert.org.
5 'Hubble Reveals Dynamic Atmospheres of Uranus, Neptune' (7 February 2019), www.nasa.gov.
6 Ibid.
7 Bradford Smith et al., 'Voyager 2 in the Uranian System: Imaging Science Results', *Science*, CCXXXIII/4759 (4 July 1986), p. 45.
8 Walter Wild et al., 'Observation and Interpretation of a Near Infrared Spot Feature on Uranus', *Bulletin of the American Astronomical Society*, XXV/18.10 (1993), p. 1078.
9 'Hubble Discovers Dark Cloud in the Atmosphere of Uranus' (28 September 2006), www.hubblesite.org.
10 Raymond LeBeau et al., 'A Numerical Investigation of the Berg Feature on Uranus as a Vortex-Driven System', *Atmosphere*, XI/1 (2020), p. 52.
11 Imke de Pater et al., 'Post-Equinox Observations of Uranus: Berg's Evolution, Vertical Structure, and Track towards the Equator', *Icarus*, CCXV/1 (September 2011), pp. 332–45.
12 Michael Kramer, 'Giant Storms Are Raging on Uranus' (12 August 2014), www.space.com.
13 William Dunn et al., 'A Low Signal Detection of X-Rays from Uranus', *JGR Space Physics*, CXXVI/4 (31 March 2021), pp. 2–10.

14 Robin M. Canup and William R. Ward, 'A Common Mass Scaling for Satellite Systems of Gaseous Planets', *Nature*, CDXLI (2006), pp. 834–9.
15 Shigeru Ida et al., 'Uranian Satellite Formation by Evolution of a Water Vapour Disk Generated by a Giant Impact', *Nature Astronomy*, IV (2020), pp. 880–85.
16 Ibid.
17 Leigh N. Fletcher et al., 'Ice Giant Systems: The Scientific Potential of Orbital Missions to Uranus and Neptune', *Planetary and Space Science* (8 June 2020).
18 See 'Overview: Pluto – NASA Solar System Exploration', https://solarsystem.nasa.gov, accessed 12 April 2022.
19 Ali M. Bramson et al., 'Oceanus: A Uranus Orbiter Concept Study from the 2016 NASA/JPL Planetary Science Summer School', 48th Lunar and Planetary Science Conference (2017).
20 Michael Schirber, 'Mission to Mysterious Uranus' (12 October 2011), www.phys.org.
21 Tatiana Bocanegra-Bahamóna et al., 'Mission to the Uranus System: MUSE, Unveiling the Evolution and Formation of Icy Giants', Post Alpbach Summer School, Madrid, Spain (June 2012). PDF accessible at https://mlaneuville.github.io/papers/Bocanegra+2015.pdf.

4 Moons of Uranus

1 Denison Olmsted, *Introduction to Astronomy* (New York, 1839), p. 205.
2 William Lassell, 'On the Interior Satellites of Uranus', extract of a letter dated 13 November 1851.
3 David W. Hughes, 'The Historical Unravelling of the Diameters of the First Four Asteroids', *Quarterly Journal of the Royal Astronomical Society*, XXV (September 1994), p. 331.
4 Robert Stawell Ball, *Star-Land: Being Talks with Young People about the Wonders of the Heavens* (Bosto, MA, 1899).
5 Arthur Francis O'Donel Alexander, *The Planet Uranus: A History of Observation, Theory and Discovery* (New York, 1965), p. 137.
6 Ibid.
7 William H. Steavenson, 'Observations of the Satellites of Uranus', *Monthly Notices of the Royal Astronomical Society*, CVIII/2 (April 1948), pp. 183–5.
8 William Shakespeare, *The Tempest* (London, 1611).
9 Ellis D. Miner, *Uranus: The Planet, Rings and Satellites* (London, 1990).
10 Paul Helfenstein et al., 'Oberon: Color Photometry and Its Geological Implications', *Abstracts of the Lunar and Planetary Science Conference*, XXI (1990), p. 498.

11 Bradford A. Smith et al., 'Voyager 2 in the Uranian System: Imaging Science Results', *Science*, CCXXXIII/4759 (4 July 1986), pp. 43–64.
12 'Gertrude on Titania', *Gazetteer of Planetary Nomenclature*, USGS Astrogeology (1 October 2006), available at https://planetarynames.wr.usgs.gov/Feature/2150, accessed 20 March 2021.
13 Garry Hunt and Patrick Moore, *Atlas of Uranus* (Cambridge, 1989), p. 81.
14 Ibid., p. 83.
15 Paul M. Schenk, 'Fluid Volcanism on Miranda and Ariel: Flow Morphology and Composition', *Journal of Geophysical Research*, XCVI/1887 (1991), pp. 1887–1906.
16 Steve K. Croft and L. A. Soderblom, 'Geology of the Uranian Satellites', in *Uranus*, ed. Jay T. Bergstralh, Ellis D. Miner and Mildred Shapley Matthews (Tucson, AZ, 1991), pp. 693–735.
17 Miner, *Uranus*, p. 311.
18 Peter C. Thomas,'Radii, Shapes, and Topography of the Satellites of Uranus from Limb Coordinates', *Icarus*, LXXIII/3 (1988), pp. 427–41.
19 'Uranus A-buzz with Moonlets' (20 September 1999), www.spacedaily.com.

5 Neptune, Its Discovery and Pre-Voyager Observations

1 William Sheehan et al., *Neptune: From Grand Discovery to a World Revealed: Essays on the 200th Anniversary of the Birth of John Couch Adams* (London, 2021).
2 George Biddell Airy, letter from Airy to Challis, dated 9 July 1846.
3 Robert J. Mann, 'The Recent Progress of Astronomy', reprinted from *Littell's Living Age in Edinburgh Review*, CLXIX (1886), p. 460.
4 John Couch Adams, 'On the Perturbations of Uranus', *Appendices to Various Nautical Almanacs* (London,between 1834 and 1854), p. 265.
5 Hiu Man Lai, C. C. Lam and Kenneth Young, 'Perturbation of Uranus by Neptune: A Modern Perspective', *American Journal of Physics*, LVIII (1990), pp. 946–53.
6 Richard Baum and William Sheehan, *In Search of Planet Vulcan* (London, 1997), p. 108.
7 Urbain Jean Joseph Le Verrier, letter from Le Verrier to Airy, dated 26 February 1847.
8 Stillman Drake and Charles T. Kowal, 'Galileo's Observations of Neptune', *Nature*, CCLXXXVII (1980), pp. 311–13.
9 William Lassell, letter to *The Times* newspaper, London, dated 14 October 1846.
10 Richard Baum, *The Planets: Some Myths and Realities* (Frome, 1973), pp. 141–6.
11 Arthur Adel and Vesta M. Slipher, 'On the Identification of the Methane Bands in the Solar Spectra of the Major Planets', *Physical Review Journal*, XLVI/240 (August 1934). pp. 240–41.

12 Edward Emerson Barnard,'Observations of the Diameter of Neptune and of the Position of His Satellite', *Astronomical Journal*, XV/342 (1895), pp. 41–4.
13 Edward Emerson Barnard, 'Measures of the Satellite of Neptune with the 40-Inch Refractor of the Yerkes Observatory, with Remarks on the Great Telescope', *Astronomical Journal*, XIX/436 (1898), pp. 25–9.
14 Patrick Moore, *The Planet Neptune* (Chichester, 1988), p. 53.
15 Mann, ''The Recent Progress of Astronomy', reprinted from *Littell's Living Age in Edinburgh Review*, p. 462.
16 Harold Jeffreys, 'The Constitution of the Four Outer Planets', *Monthly Notices of the Royal Astronomical Society*, LXXXIII/6 (April 1923), p. 350.
17 Robert Wildt, 'Reports on the Progress of Astronomy: The Constitution of the Planets', *Monthly Notices of the Royal Astronomical Society*, CVII (March 1947), p. 84.
18 William H. Ramsey, 'On the Constitutions of the Major Planets', *Monthly Notices of the Royal Astronomical Society*, CXI (May 1951), p. 427.
19 George Wesley Lockwood, 'Planetary Brightness Changes: Evidence for Solar Variability', *Science*, CXC/560 (1975), pp. 560–62.
20 Richard R. Joyce et al., 'Evidence for Weather on Neptune', *Astrophysics Journal*, CCXIV (1977), pp. 657–62.
21 Bradford A. Smith and Harold J. Reitsema, 'Imaging of Uranus and Neptune', Department of Planetary Sciences and Lunar and Planetary Laboratory, University of Arizona (1982).
22 Bradford A. Smith, 'Near Infrared Imaging of Uranus and Neptune', JPL *Uranus and Neptune* (1984), pp. 213–22.
23 Garry Hunt and Patrick Moore, *Atlas of Neptune* (Cambridge, 1994), p. 28.
24 Ibid.

6 Voyager 2 Flypast of Neptune

1 Patrick Moore, *The Planet Neptune* (Chichester, 1988), p. 7.
2 Thomas William Webb, *Celestial Objects for Common Telescopes* (London, 1917).
3 '25 Years after Neptune: Reflections on Voyager' (25 August 2014), https://voyager.jpl.nasa.gov.
4 'The Reg Chats with Voyager Imaging Team Member Dr Garry E. Hunt' (18 September 2018), www.theregister.com.
5 '25 Years after Neptune: Reflections on Voyager' (25 August 2014), https://voyager.jpl.nasa.gov.
6 George Wesley Lockwood and Mikołaj Jerzykiewicz, 'Photometric Variability of Uranus and Neptune, 1950–2004', *Icarus*, CLXXX (2006), pp. 442–5.
7 Ed C. Stone and Ellis D. Miner, 'The Voyager 2 Encounter with the Neptune System', *Science*, CCXLVI/4936 (15 December 1989), pp. 1417–21.

8 Ed C. Stone, 'Voyager through the Eyes of Project Scientist Ed Stone', *Astronomy* (2014), www.astronomy.com/bonus/edstone.
9 Ellis D. Miner and Randii R. Wessen, *Neptune: The Planet, Rings and Satellites* (London, 2002), p. 238.
10 Linda J. Horn et al., 'Observations of Neptunian Rings by Voyager Photopolarmeter Experiment', *Geophysical Research Letters*, XVII/10 (September 1990), pp. 1745–8.
11 Carolyn Porco, 'An Explanation for Neptune's Ring Arcs', *Science*, CCLIII/5023 (August 1991), pp. 995–1001.
12 Faith Namouni and Carolyn Porco, 'The Confinement of Neptune's Ring Arcs by the Moon Galatea', *Nature*, CDXVII (2002), pp. 45–7.
13 Bruno Sicardy et al., 'Neptune's Rings, 1983–1989: Ground-Based Stellar Occultation Observations I, Ring-Like Arc Detections', *Icarus*, LXXXIX/2 (February 1991), pp. 220–43.
14 Ellis D. Miner, Randii R. Wessen and Jeffrey N. Cuzzi, *Planetary Ring Systems* (London, 2007), pp. 69–71.
15 Francis Reddy and Ed Stone, 'Voyager through the Eyes of Project Scientist Ed Stone', *Astronomy: Voyager 40th Anniversary Edition*, bonus online content available at https://astronomy.com/bonus/edstone (August 2017).
16 'Where Are the Voyagers Right Now?', https://voyager.jpl.nasa.gov, accessed March 2021.

7 Neptune after Voyager

1 William B. Hubbard and Joseph J. MacFarlane, 'Structure and Evolution of Uranus and Neptune', *Journal of Geophysical Research*, LXXXVIII (January 1980), p. 225.
2 Martin Ross, 'The Ice Layer in Uranus and Neptune – Diamonds in the Sky?', *Nature*, CCXCII (1981), pp. 435–6.
3 'Neptune's Stormy Disposition' (21 May 1999), www.jpl.nasa.gov.
4 'Hubble Tracks the Lifecycle of Giant Storms on Neptune' (25 March 2019), www.nasa.gov.
5 Ibid.
6 Robert Sanders, '"Dark Vortex" Confirmed on Neptune' (23 June 2016), www.vcresearch.berkeley.edu.
7 Ibid.
8 'Dark Storm on Neptune Reverses Direction, Possibly Shedding a Fragment' (2020), www.nasa.gov.
9 'New Storm Makes Surprise Appearance on Neptune' (2 August 2017), www.keckobservatory.org.

10 Ibid.
11 Ibid.
12 Hannah Wakeford et al., 'HAT-P-26b: A Neptune-Mass Exoplanet with a Well-Constrained Heavy Element Abundance', *Science*, CCCLVI/6338 (12 May 2017), pp. 628–31.
13 Heather A. Knutson et al., 'A Spitzer Transmission Spectrum for the Exoplanet GJ 436b, Evidence for Stellar Variability, and Constraints on Dayside Flux Variations', *Astrophysical Journal*, DCCXXXV/27 (July 2011), p. 23.
14 David Ehrenreich et al., 'A Giant Comet-Like Cloud of Hydrogen Escaping the Warm Neptune-Mass Exoplanet GJ 436b', *Nature*, DXXII (2015), pp. 459–61.
15 James S. Jenkins et al., 'An Ultra-Hot Neptune in the Neptune Desert', *Nature Astronomy*, IV (21 September 2020), p. 1202.
16 Olivier Mousis et al., 'Irradiated Ocean Planets Bridge Super-Earth and Sub-Neptune Populations', *Astrophysical Journal Letters*, DCCCXCVI/2 (15 June 2020), pp. 896, L22.

8 Moons of Neptune

1 William Lassell, 'Observations of Neptune and Its Satellite Triton', *Monthly Notices of the Royal Astronomical Society*, VII (1847), p. 307. Italics are in the original.
2 Henry Norris Russell, 'Mass Density and Name of Triton', *Nature*, CXXIX (1932), p. 405.
3 Raymond Arthur Lyttleton, 'The Origin of the Solar System', *Monthly Notices of the Royal Astronomical Society*, XCVI (1936), pp. 559–68.
4 George Ellery Hale, 'Mount Wilson Observatory', *Mount Wilson Observatory Annual Report*, III (1931), pp. 171–226.
5 Garry Hunt and Patrick Moore, *Atlas of Neptune* (Cambridge, 1994), p. 71.
6 Harold J. Reitsema et al., 'Occultation by a Possible Third Satellite of Neptune', *Science*, CCXV/4530 (1982), pp. 289–91.
7 Ben Evans, 'Remembering Voyager 2's Visit with Neptune, 30 Years On (Part 2)' (18 August 2019), www.americaspace.com.
8 Alan Stern and William B. McKinnon, 'Triton's Surface Age and Impactor Population Revisited (Evidence for an Internal Ocean)', *Lunar and Planetary Science XXX: Papers Presented at the Thirtieth Lunar and Planetary Science Conference 15–19 March, 1999* (Washington, DC, 1999).
9 Francis Reddy and Ed Stone, 'Voyager through the Eyes of Project Scientist Ed Stone', *Astronomy: Voyager 40th Anniversary Edition* (August 2017).
10 Stern and McKinnon, 'Triton's Surface Age', p. 945.
11 Ellis D. Miner and Randii R. Wessen, *Neptune: The Planet, Rings and Satellites* (London, 2002), p. 260.

12 'MIT Researcher Finds Evidence of Global Warming on Neptune's Largest Moon' (24 June 1998), https://news.mit.edu.
13 Hunt and Moore, *Atlas of Neptune*, p. 59.
14 Csaba Kiss et al., 'Nereid from Space: Rotation, Size and Shape Analysis from K2, Herschel and Spitzer Observations', *Monthly Notices of the Royal Astronomical Society*, CDLVII/3 (2016), pp. 2908–17.
15 Michael E. Brown et al., 'Detection of Water Ice on Nereid', *Astrophysical Journal*, DVIII (1998), pp. L175–L176.
16 Mark Showalter, Ray Villard and Jack Lissauer, 'Tiny Neptune Moon May Have Broken from Larger Moon' (20 February 2019), https://solarsystem.nasa.gov.
17 'Proposed NASA Mission Would Visit Neptune's Curious Moon Triton' (16 June 2020), www.nasa.gov.
18 Ibid.
19 Steven Oleson and Geoffrey Landis, 'Triton Hopper: Exploring Neptune's Captured Kuiper Belt Object', Planetary Science Vision 2050 Workshop 2017. PDF can be accessed at www.hou.usra.edu.

9 Observing the Ice Giants

1 Martin Stangl, personal correspondence with author via email, November 2020.
2 Richard W. Schmude Jr et al., 'The Secular and Rotational Brightness Variations of Neptune' (2016), https://ui.adsabs.harvard.ed, accessed March 2021.
3 Mike Foulkes, personal correspondence with author via email, April 2021.

Bibliography

Alexander, Arthur Francis O'Donel, *The Planet Uranus: A History of Observation, Theory and Discovery* (New York, 1965)
Baum, Richard, *The Planets: Some Myths and Realities* (Frome, 1973)
—, and William Sheehan, *In Search of Planet Vulcan* (London, 1997)
Cruikshank, Dale P., and William Sheehan, *Discovering Pluto: Exploration at the Edge of the Solar System* (Tucson, AZ, 2019)
Hunt, Garry, and Patrick Moore, *Atlas of Neptune* (Cambridge, 1994)
—, *Atlas of Uranus* (Cambridge, 1989)
Littman, Mark, *Planets Beyond: Discovering the Outer Solar System* (New York, 1990)
Mack, Pamela E., ed., *From Engineering Science to Big Science: The NACA and NASA Collier Trophy Research Project Winners* (Washington, DC, 1998)
Miner, Ellis D., *Uranus: The Planet, Rings and Satellites* (London, 1990)
—, and Randii R. Wessen, *Neptune: The Planet, Rings and Satellites* (London, 2002)
—, — and Jeffrey N. Cuzzi, *Planetary Ring Systems* (London, 2007)
Moore, Patrick, *New Guide to the Planets* (London, 1993)
—, *The Planet Neptune* (Chichester, 1988)
Rothery, David A., *Satellites of the Outer Planets* (Oxford, 1992)
Sheehan, William, et al., *Neptune: From Grand Discovery to a World Revealed: Essays on the 200th Anniversary of the Birth of John Couch Adams* (London, 2021)
Stern, Alan, and David Grinspoon, *Chasing New Horizons: Inside the Epic First Mission to Pluto* (London, 2018)

Acknowledgements

I wish to express my gratitude and thanks to, first of all, Peter Morris for his extensive help and comments, and for making this a better book, as well as William Sheehan for his guidance, and Mike Foulkes and Kevin Kilburn for their very insightful read-throughs. For use of their images: Damian Peach, Martin Stangl and Rolf Winkler, Professor Imke de Pater and the Keck Observator and Dr Sian Prosser, Librarian at the Royal Astronomical Society. As always, I wish to thank Brian Sheen from the Roseland Observatory. I would also like to thank Jamie Ashley and my family for their love and patience as this book was written.

Photo Acknowledgements

The author and publishers wish to thank the organizations and individuals listed below for authorizing reproduction of their work.

ESA/Hubble Space Telescope: p. 144; ESA/Hubble & NASA, L. Lamy/Observatoire de Paris: p. 63; ESO: p. 93; ESO/L. Calçada: p. 172; ESO/P. Weilbacher (AIP): p. 132; From *First Steps in Astronomy and Geography* (London, 1828): p. 76; Linda Hall Library, Kansas City, Missouri: p. 19; W. Herschel, *Scientific Papers*: pp. 15 (vol. 1, p. 33), 24 (vol. 1 p. 314); from *Hutchinson's Splendour of the Heavens*: p. 29 (Phillips, Theodore Evelyn Reece and Steavenson, William Herbert, vol. 1, p. 378, 1923); Edward Molter and Imke de Pater, University of California, Berkeley: p. 56, Edward Molter and Imke de Pater, University of California, Berkeley/C. Alvarez, W. M. Keck Observatory: p. 140; NASA: pp. 34, 36, 69, 70, 171; NASA/CXO/University College London/W. Dunn et al.: p. 65; NASA, ESA and A. Field (STSCI): pp. 145, 167 top; NASA, ESA and M. Showalter (SETI Institute): pp. 54, 57 left and right, 167 bottom, 168; NASA, ESA, A. Simon (NASA Goddard Space Flight Center), and M. H. Wong and A. Hsu (University of California, Berkeley): pp. 9 top and centre, 62; NASA, ESA, L. Sromovsky and P. Fry (University of Wisconsin), H. Hammel (Space Science Institute) and K. Rages (SETI Institute): p. 59; NASA, ESA, and M. H. Wong and A. I. Hsu (University of California, Berkeley): p. 138; NASA, ESA, STSCI, M. H. Wong (University of California, Berkeley) and L. A. Sromovsky and P. M. Fry (University of Wisconsin–Madison): p. 139; NASA, ESA, and M. H. Wong and J. Tollefson (University of California, Berkeley): p. 137; NASA/GSFC: p. 143; NASA/Hubble Space Telescope: pp. 85 bottom, 91; NASA/JPL: pp. 42, 47 bottom, 50, 52, 81, 82, 83, 84, 85 top, 87 top and bottom, 88, 89 top and bottom, 90 top, 117, 118, 119 top and bottom, 121, 126, 127, 129, 153, 155 left and right, 157, 159, 160, 161, 162 top and bottom, 163, 164 left; NASA/JPL-Caltech: pp. 8 top, centre and bottom, 37, 38, 47 top, 170; NASA/JPL-Caltech, Lawrence Sromovsky, University of Wisconsin–Madison/W. W. Keck Observatory: p. 61;

NASA/JPL/Universities Space Research Association/Lunar and Planetary Institute: p. 158; NASA/JPL/STSCI: pp. 135, 136; NASA/JPL/United States Geological Survey: p. 152; NASA's Scientific Visualization Studio – JPL NAIF: p. 123 (image extracted from a video. Tom Bridgman (GST); Lead Animator Mara Johnson-Groh (Wyle Information Systems); Writer Laurence Schuler (ADNET); Project Support Ian Jones (ADNET). Project Support Data Used: Luhmann-Friesen Magnetosphere Model (1979)/Magnetic Field Lines (Luhmann-Friesen) DE 430); Damien Peach: pp. 174 top right, 177 centre left; Royal Astronomical Society Library: pp. 28, 98 (MSS Add. 94, item 24), 100 left (MSS Add. 91. Astronomical scrapbook of Mrs Hannah Jackson-Gwilt, item 98a), 100 right (MSS Add. 91. Astronomical scrapbook of Mrs Hannah Jackson-Gwilt, item 98b); Kenneth Seidelmann, U.S. Naval Observatory/NASA: p. 90 bottom; Smithsonian Museum: p. 104 (*Smith's Illustrated Astronomy*, Asa Smith, 1849); Martin Stangl: pp. 174 top left, 175; Martin Stangl and Rolf Winkler: p. 177 top left; from John Timbs, *The Year-Book of Facts in Science and Art* (London, 1846): p. 73; United States Naval Observatory: p. 79; VLT, ESA, NASA, JPL and Paris Observatory: p. 133; Yale Center for British Art: p. 14.

AlfvanBeem, the copyright holder of the image on p. 16, has published it online under conditions imposed by a Creative Commons CC0 1.0 Universal Public Domain Dedication License. LittlefieldEmery, the copyright holder of the image on p. 30, has published it online under conditions imposed by a Creative Commons Attribution-Share Alike 4.0 International License. FrancescoA, the author of the work on p. 49, has released it into the public domain. RJHall, the copyright holder of the image on p. 97 right, Cédric Foellmi, the copyright holder of the image on p. 113, and Fama Clamosa, the copyright holder of the image on p. 164 right, have published them online under conditions imposed by a Creative Commons Attribution-Share Alike 3.0 Unported License; User:Eurocommuter, the copyright holder of the image on p. 165, has published it online under conditions imposed by a Creative Commons Attribution-Share Alike 3.0 Unported License. (Plotted by a program written by the User:Eurocommuter, Mean orbital parameters from JPL, February 2007.); Public Domain: pp. 75, 97 left, 101 top and centre, 109; Wikiwand (Public Domain): p. 45.

Index

Page numbers in ***bold italics*** refer to illustrations